AF307212

Current Topics in Microbiology and Immunology

180

Pathogenesis of Shigellosis

Edited by P. J. Sansonetti

With 15 Figures

Springer-Verlag
Berlin Heidelberg NewYork
London Paris Tokyo
Hong Kong Barcelona
Budapest

P. J. SANSONETTI, M.D.
Institut Pasteur
Unité de Pathogénie Microbienne Moleculaire
28, Rue du Dr. Roux
75724 Paris Cedex 15
France

Cover illustration: Movement of *Shigella flexneri* within Hela cells. Actin bundles forming two protrusions are observed in scanning electron microscopy after fixation that insolubilizes the cell cytoskeleton. Bacteria can be seen at the tip of these protrusions.
M. Lesourd, P. J. Sansonetti, Station Centrale de Microscopie Electronique, Institut Pasteur.

ISBN-13:978-3-642-77240-5 e-ISBN-13:978-3-642-77238-2
DOI: 10.1007/978-3-642-77238-2

Softcover reprint of the hardcover 1st edititon 1992

Library of Congress Catalog Card Number 15-12910

23/3020-5 4 3 2 1 0 – Printed on acid free paper.

Preface

Shigellosis is present all over the world. Anyone traveling in developing countries knows that the control of this invasive disease of the intestine is a priority task for physicians and public health authorities. Victims are essentially young children, and complications such as the hemolytic uremic syndrome make shigellosis a systemic disease rather than simply an infection of the colonic mucosa. However, "Westerners" should not consider shigellosis as an unlikely threat of the tropics. The disease arises in industrialized countries as soon as breaches in sanitation appear. A few months ago, at least 500 people developed shigellosis in northern France in an outbreak of *Shigella sonnei* infection due to accidental contamination of an urban water delivery system.

The pathogenesis of shigellosis is an extraordinary topic of research because study of the invasion of the colonic mucosa addresses fundamental questions on the molecular and cellular mechanisms by which a bacterial pathogen can penetrate nonphagocytic cells, survive, multiply, spread in the intracellular compartment, and eventually kill host cells. Further development of the infection within subepithelial tissues as well as the mechanisms that contribute to the eradication of this process have barely been studied.

Therefore, the questions that are addressed in this system are of general interest for those comparing the various strategies that bacteria, viruses, parasites, and fungi have developed to invade human and animal mucosal surfaces. On the other hand, all the authors in this volume are aware that contributing to the development of tools that will contribute to the control of the disease is an aim of their research.

This volume is an up-to-date compilation of information on the pathogenesis of shigellosis. Within the limits of its format, it has been possible to address essentially all recent developments in the molecular and cellular basis of epithelial cell invasion, the genetic, biochemical, and molecular mode of action of endotoxin (lipopolysaccharide) and exotoxin (Shiga toxin), and the emerging field of environmental regulation of virulence. Last but not least, all this information has been synthesized as a logical introduction to vaccine development. This exhaustive coverage is the result of a tremendous collaborative effort among some of the major researchers in this field. Many of them have agreed to contribute to the preparation of individual chapters under the supervision of a coeditor, and I wish to express our thanks to all of them for having made this adventure a real success.

Finally, all the contributors to this volume have been influenced, to varying degrees, by a man who has been a pioneer in this field as well as a model of behavior in science. I would like to dedicate this book to Sam B. Formal.

P. J. Sansonetti

List of Contents

List of Contributors

(Their addresses can be found at the beginning of the respective chapters)

Molecular and Cellular Biology of *Shigella flexneri* Invasiveness: From Cell Assay Systems to Shigellosis

P. J. Sansonetti

1 Introduction

Also known as bacillary dysentery, shigellosis, an invasive disease of the human colon, is present worldwide. It is of particular concern in tropical regions, especially in overcrowded areas of the developing world, where *Shigella flexneri* causes the endemic form of the disease and *Shigella dysenteriae* 1 devastating epidemics. Children are the most common victims of this disease which is principally transmitted by the fecal-oral route in areas having inadequate hygiene and sanitation.

Shigellosis is characterized by fever, abdominal cramps, tenesmus, and dysenteric stools containing mucus, blood, and pus. These symptoms reflect invasion of the colonic mucosa by *Shigella*, the critical stage in the pathogenic process (LABREC et al. 1964). Invasion occurs in two steps: invasion of the cells that form the epithelial lining of the colon (TAKEUCHI et al. 1965) and invasion

Unité de Pathogénie Microbienne Moléculaire, INSERM U199, Institut Pasteur, 28 Rue du Dr. Roux, 75724 Paris Cédex 15, France

Current Topics in Microbiology and Immunology, Vol. 180

of the connective tissue (lamina propria) of the intestinal villi. The invasive process elicits a strong inflammatory response, leading to focal abscess formation and ulceration (TAKEUCHI et al. 1968). In addition, systemic complications, such as the hemolytic uremic syndrome (HUS), are observed in a few cases (KOSTER et al. 1978). Shiga toxin, a potent cytotoxin produced in high quantity by *S. dysenteriae* 1, may have a role in the pathogenesis of HUS, but this is still being debated (O'BRIEN and HOLMES 1987). Its precise role as an adjuvant to the invasive process also needs further evaluation (FONTAINE et al. 1988; NEILL et al. 1988).

In vitro, cell assay systems allow examination of the molecular and cellular basis of cell invasion. Several aspects can be explored: (a) entry of the bacteria into cells that are nonprofessional phagocytes; (b) intracellular multiplication; (c) intracellular and cell-to-cell spread of the bacteria; and eventually (d) host cell killing. The relevance of in vitro studies must be assessed by reverse genetics and subsequent testing of mutants in more definitive virulence assays, such as the guinea pig keratoconjunctivitis assay or Sereny test (SERENY et al. 1957), the rabbit ligated ileal loop assay (GOTS et al. 1974), or intragastric infection of rhesus monkeys which develop a dysenteric syndrome resembling shigellosis (TAKEUCHI et al. 1968). This paper will review our current knowledge on the molecular and cellular basis of cell invasion by *S. flexneri*. When possible, the in vivo relevance of these data will be addressed.

2 Entry of *S. flexneri* into Epithelial Cells

2.1 Nature of the Entry Process

As shown in Fig. 1, transmission electron microscopy (TEM) performed on epithelial cell monolayers (HeLa cells) demonstrates that invasive isolates of *S.flexneri* are internalized via an endocytic process that involves formation of long pseudopods, at times with anarchic alterations of the cell surface. This process is active, requiring energy produced by both the bacterium and the host cell (HALE et al. 1979), and is inhibited by cytochalasin B and D, molecules that cap the growing end of polymerizing actin filaments (TANENBAUM 1978). These data provide indirect evidence that *S. flexneri* is internalized via bacterium-directed phagocytosis. In professional phagocytes such as polymorphonuclear leucocytes and monocyte/macrophages, actin polymerized in microfilaments together with actin-binding proteins such as myosin forms a molecular complex that is the driving force of the phagocytic process (STENDAHL et al. 1980; SHETERLINE et al. 1984). During the phagocytic process triggered by *S. flexneri*, the role of the cellular cytoskeletal microfilaments can be studied directly by microscopy, using fluorescent probes such as 7-nitrobenz-2-oxa-1,3-diazole (NBD)-phallacidin (BARAK et al. 1980), which binds specifically to filamentous

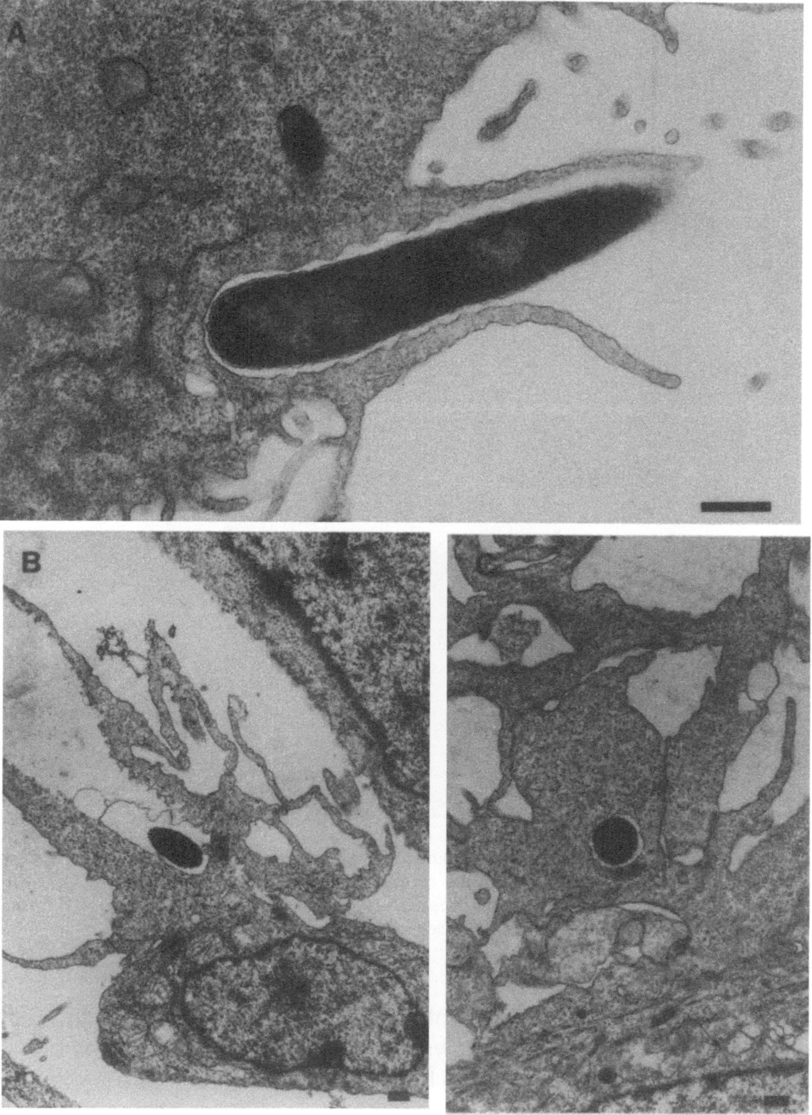

Fig. 1 A–C. TEM demonstrating the entry process of *S. flexneri* into HeLa cells via directed phagocytosis. **A** Two pseudopods engulfing the invading microoraganism. **B, C** Single microorganism inducing anarchic polymerization of actin filaments located beneath the cytoplasmic membrane. *Bars*, 0.5 μm

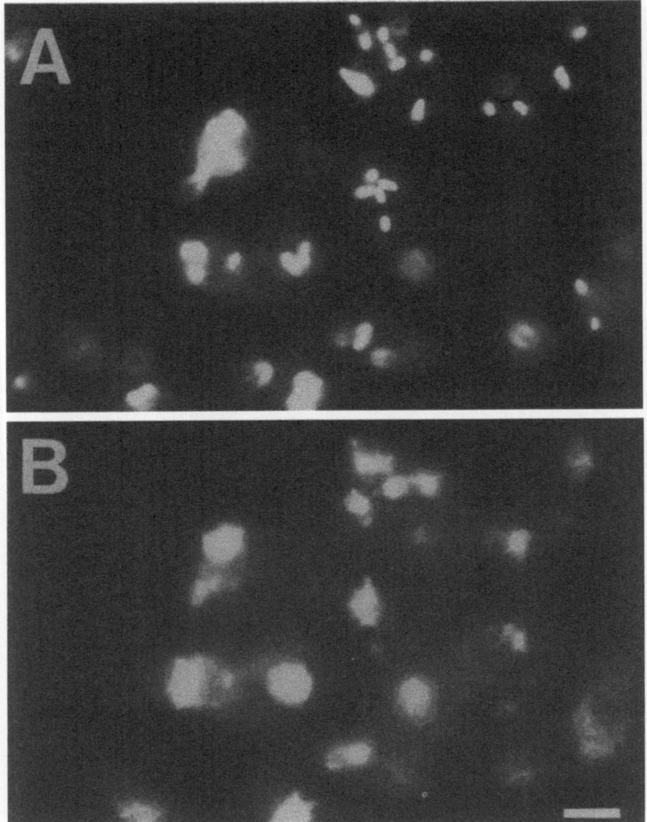

Fig. 2 A, B. Double immunofluorescence labeling of HeLa cells infected by an invasive isolate of *S. flexneri.* **A** Labeling of bacteria. Incubation of infected HeLa cells with rabbit antilipopolysaccharide serum followed by goat rhodamine-labeled, anti-rabbit immunoglobulin serum. **B** Labeling of polymerized actin (F-actin) by NBD-phallacidin. *Bar*, 10 μm

actin (F-actin), and monoclonal antibodies directed against actin-binding proteins, particularly myosin. As shown in Fig. 2, F-actin accumulates beneath the plasma membrane in areas of the HeLa cell surface that interact with strain M90T, an invasive isolate of *S. flexneri* serotype 5a. Similar observations have been made for myosin (CLERC and SANSONETTI 1987). Measurement of the monomeric versus total actin content of synchronously infected HeLa cells suggests transient de novo polymerization of the pool of monomeric actin (G-actin) with maximal polymerization occurring as soon as 6 min after initial contact of the microorganism with the cell surface. A noninvasive mutant of M90T, BS176, does not elicit significant modifications of cellular microfilaments. These data indicate that *S. flexneri* has the capacity to induce epithelial cells to perform a phagocytic process very similar to that performed by professional phagocytes.

2.2 Bacterial Genes and Protein Products Responsible for the Entry Process

Presence of a large 220-kb plasmid is necessary for expression of the invasive phenotype of *S. flexneri* (SANSONETTI et al. 1982, 1983), as well as of other *Shigella* species (reviewed by SASAKAWA et al., this volume). We will concentrate here only on the genes and protein products involved in the entry process. The first successful attempt at cloning the genes involved in entry was performed in strain M90T. It required the use of a cosmid vector and led to the identification of a 37-kb segment on the large plasmid (MAURELLI et al. 1985). Subsequent analysis by Tn5 mutagenesis of one of the recombinant cosmids, pHS4108 (BAUDRY et al. 1987), and of the whole virulence plasmid of a *S. flexneri* serotype 2a isolate (SASAKAWA et al. 1988) identified five independent contiguous loci within a 30-kb segment. This *inv* region is still incompletely characterized; studies are underway to identify the numerous genes and to characterize the role of their respective protein products in the entry process. Locus 1 is known to contain a positive regulatory gene named *virB* in *S. flexneri* serotype 2a (ADLER et al. 1989), *ipaR* in *S. flexneri* serotype 5a (BUYSSE et al. 1990), and *invE* in *Shigella sonnei* (WATANABE et al. 1990). Under the control of *virF*, a second positive regulatory gene, *virB* (SAKAI et al. 1986), induces several invasion genes contained in other loci. The most extensively studied of these loci is locus 2, which is shown in Fig. 3. It contains the genes encoding four proteins that are highly immunogenic as demonstrated by the presence of specific antibody in sera from both human beings and monkeys convalescing from shigellosis (BAUDRY et al. 1987). These proteins are called Ipa for *i*nvasion *p*lasmid *a*ntigens (HALE et al. 1985; OAKS et al. 1986). The genes encoding the four Ipa are located in locus 2. Figure 3 summarizes much research on the organization of this locus (BAUDRY et al. 1988; BUYSSE et al. 1987; SASAKAWA et al. 1989). The four *ipa* genes are arranged in an operon in the following order from 5′ to 3′: *ipaB*, *C*, *D*, and *A*. Two additional genes, encoding polypeptides of 17 and 21 kDa that are not significantly immunogenic during natural infection, are also part of this operon and are located upstream of *ipaB*.

We have investigated the respective roles of these Ipa in the invasion process (HIGH and SANSONETTI, manuscript submitted). Mutations have been generated via double allelic exchange of the wild-type plasmid gene with its in vitro mutagenized counterpart. Omega, the gene cassette encoding spectinomycin resistance (PRENTKI and KIRSCH 1984), was used to select the appropriate

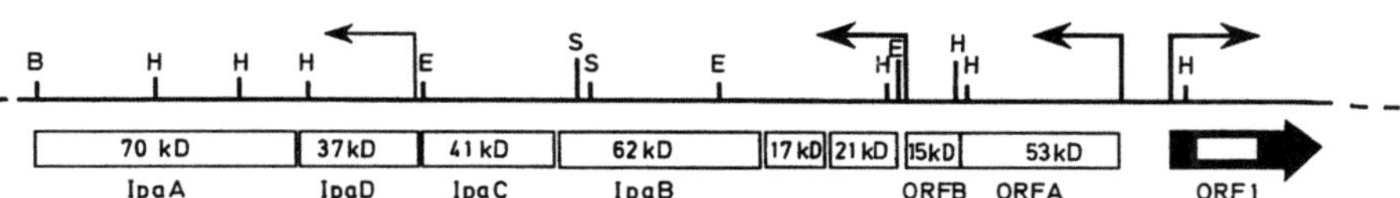

Fig. 3. Genetic map of locus 2 showing location of the *ipa* genes

mutants. *ipaB* and ipaC mutants are not invasive in the HeLa cell assay system, thus suggesting that the polypeptides they encode are "invasins" of *S. flexneri.* Although these mutants do not trigger significant polymerization of actin, they are highly adhesive to the surface of the cells, providing good evidence that an entry step can be differentiated from an adhesion step. The "adhesin" of *S. flexneri* has yet to be characterized. IpaD may be the expected adhesin or a protein required for proper organization of the adhesion/entry complex on the surface of the bacterium. Finally, *ipaA* mutants remain invasive. They elicit a positive plaque assay and are positive in the Sereny test, although initiation of keratoconjunctivitis is delayed as compared with the wild-type strain.

The organization of genes within loci 3, 4, and 5, and the function of their respective proteins are currently under study by several groups. The *mxi* genes of locus 4 and locus 5 may be involved in the proper localization of Ipa proteins (HROMOCKYJ and MAURELLI 1989).

2.3 Nature of the Host Cell Machinery That Carries Out the *Shigella*-Directed Phagocytic Process

In HeLa cells (P.J. SANSONETTI, unpublished observation) as well as in chick embryo fibroblasts (VASSELON et al. 1991), bacteria penetrate at the site of the cell adhesion plaques where cells adhere to their matrix components. This region of the cell is particularly rich in converging filaments of actin and integrins. The latter are dimeric integral transmembrane proteins that have a capacity to bind at their extracellular domain to cell matrix proteins and at their cytoplasmic domain to actin filaments via proteins such as talin and vinculin (RUOSLAHTI and PIERSCHBACHER 1987). Evidence is accumulating that certain members of the large family of integrins are important components of the cell receptor machinery that mediates binding and/or entry of some invasive pathogens. For example, Inv, the 103-kDa invasin of *Yersinia pseudotuberculosis*, binds to the β-1 subunit of several integrins (ISBERG and LEONG 1990), and FHA, the fibrillar hemagglutinin of *Bordetella pertussis*, mediates invasion of monocytes by binding to the CR3 receptor for complement which belongs to the integrin family (RELMAN et al. 1990).

2.4 Is the Actinomyosin Complex Alone Sufficient to Permit Entry into the Cells?

In a few instances, TEM has shown clathrin-coated pits associated with penetrating shigellae (HALE 1986). These structures seemed to be present by chance with no evidence to suggest that they were involved in the entry process. The diameter of a bacterium is much larger than that of an average coated pit.

At the site of *S. flexneri* entry into HeLa cells, we have shown (CLERC and SANSONETTI 1989) that, in addition to the accumulation of the actinomyosin complex beneath the cytoplasmic membrane, accumulation of clathrin molecules could be detected by immunofluorescence using a specific monoclonal antibody. By TEM, these molecules did not appear organized as coated pits (P. CLERC and P.J. SANSONETTI, unpublished data). When potassium depletion was obtained (a condition which is known to impair clathrin polymerization), entry of *S. flexneri* was completely abolished despite the persistence of actin polymerization. Although a nonspecific effect of potassium depletion on the function of the actinomyosin complex is likely, these results suggest that receptor-mediated endocytosis may also be involved in entry. Clathrin could play a role in the turnover and stabilization of the cytoplasmic membrane demaged by the ongoing entry process. Alternatively, it could be involved in generating the mechanical force that drives entry and the phagocytic process. Therefore, *S. flexneri* may be able to recruit all the cellular components necessary to generate an endocytic process.

2.5 Are the HeLa Cell and Chick Embryo Fibroblast Invasion Assays Relevant to the In Vivo Situation in Which Bacteria Invade Colonic Enterocytes?

The interface between the intestinal epithelium and the lumen consists of the apical surface of enterocytes, which display microvilli, forming the brush border. These microvilli contain bundles of filaments of polymerized actin cross-linked by fimbrin and villin and laterally bound to the membrane by a calmodulin/110 K protein complex. The latter protein belongs to the family of type 1 myosins (MOOSEKER and COLEMAN 1989; LOUVARD 1989). The expected target for *Shigella* in vivo therefore appears quite different from the HeLa cell surface encountered in vitro. If a bacterium needs to penetrate such a sophisticated system, it is expected to disorganize the brush border pattern in order to render the membrane-cytoskeleton complex flexible and the pool of available actin and actin-binding proteins sufficient for the phagocytic process to occur. In addition, sequestration of receptors during enterocyte differentiation and polarization of the cell may affect the localization of the receptor(s) to which shigellae bind. When the human colonic cell line Caco2 (ROUSSET 1986) is grown in such a way that cells form independent islets, shigellae do not enter the cells in the middle of the islets that have differentiated to become mature enterocytes displaying a fully differentiated brush border. Instead, they bind to the outer edge of the islet and subsequently enter peripheral cells (P.J. SANSONETTI and J. MOUNIER, unpublished data); this is shown in Fig. 4. Thus, in a cell assay system that is as close as possible to the in vivo target, *S. flexneri* does not recognize receptors on the apical pole of enterocytes, but rather a receptor expressed on the basolateral pole of the cell, which can only be "seen" at the periphery of islets. Again, this may be consistent with the receptor being an

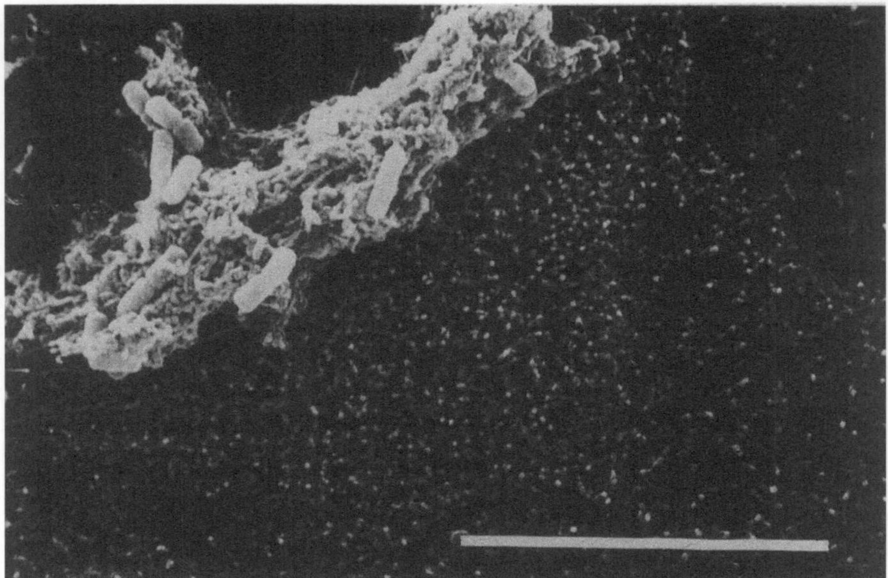

Fig. 4. Scanning electron microscopy showing initial interaction between *S. flexneri* and human colonic cells Caco-2. The adhesion and entry process occurs at the edge of the cell; microorganisms are not observed to interact with apical microvilli. *Bar*, 10 μm

integrin. Subsequently, colonization of the islet occurs by cell-to-cell spread, even if gentamicin is present in the culture medium, thereby killing extracellular bacteria. These observations have major implication for our understanding of the mechanisms of colonic invasion which will be discussed later in this review.

3 Intracellular Multiplication of *S. flexneri*

The ability of *S. flexneri* to grow intracellularly can be studied in infected HeLa cell monolayers. The generation time of strain M90T is about 40 min; five bacteria initially present in an infected cell can multiply to yield approximately 500 organisms within 4–5 h after entry (Sansonetti et al. 1986). This capacity to multiply early and efficiently within the intracellular compartment seems to be unique to shigellae. Salmonellae and yersiniae display minimal intracellular multiplication during the first 6 h of infection (Sansonetti et al. 1986). Lysis of the phagocytic vacuole surrounding shigellae occurs shortly after phagocytosis and is probably a prerequisite to efficient intracellular multiplication (Fig. 5). A virulent strain of *S. flexneri* that multiplies efficiently within HeLa cells lyses its phagocytic vacuole within 30 min of entry (Sansonetti et al. 1986). Lysis of the phagocytic vacuole releases the bacteria into the cytoplasm, thus providing better conditions for rapid bacterial growth. This is also thought to prevent phagolysosomal fusion, thus protecting bacteria against being killed by phagocytic cells. In addition, it allows more efficient host cell killing, intracellular movement, and cell-to-cell spread, as will be discussed later. Lysis of the phagocytic vacuole has been correlated to expression of a contact hemolytic activity (CLERC et al. 1986). Noninvasive strains are nonhemolytic. Virulent strains grown at 30 °C do not express the invasive phenotype and are also nonhemolytic. Similarly, Tn5 mutants in the invasion genes also inhibit the hemolytic phenotype (BAUDRY et al. 1987), which increases by 100- to 1000-fold

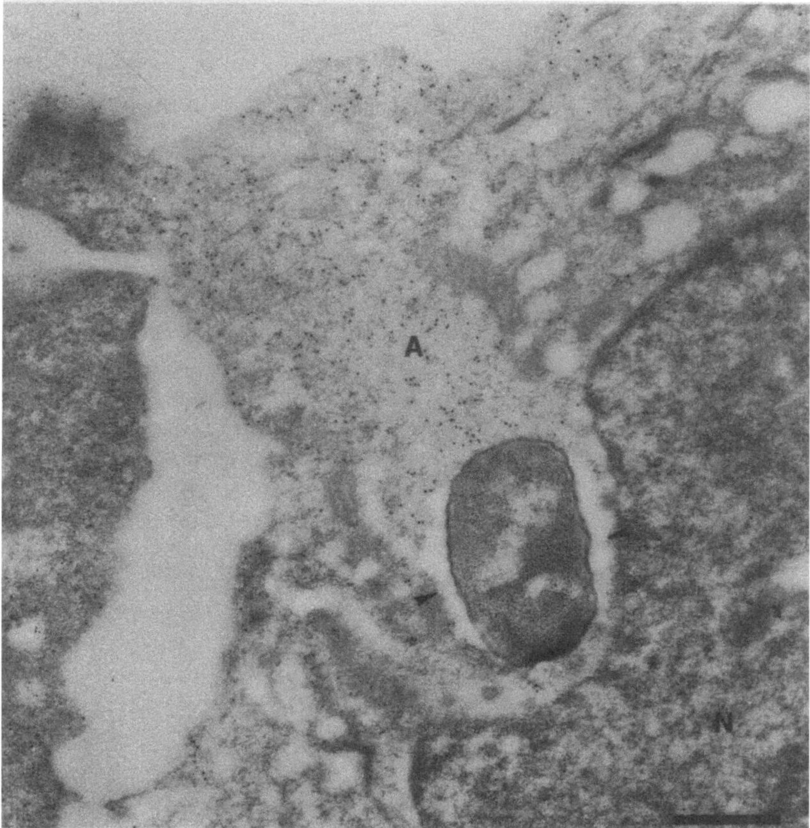

Fig. 5. TEM performed on HeLa cells infected for 2 h by *S. flexneri*. Immunogold labeling is performed with an anti-actin monoclonal antibody. This figure summarizes two major points: (a) lysis of the membrane-bound phagocytic vacuole by the invading microorganism is indicated by *two arrowheads*; (b) the presence of a gold-labeled actin tail which materializes the intracellular movement of the bacterium. Note that the bacterium repels the nuclear membrane *N*, nucleus; *A*, actin. *Bar*, 0.5 μm

as the pH drops from 7 to 5.5 (CLERC et al. 1987). This suggests that the molecular complex necessary for bacteria to enter cells also accounts for the contact hemolytic activity that occurs as the pH of the phagocytic vacuole decreases. The intracellular behavior of *S. flexneri* differs from that of other enteroinvasive genera such as salmonellae and yersiniae which do not lyse their phagocytic vacuoles (SANSONETTI et al. 1986; SMALL et al. 1987). In contrast, *Listeria monocytogenes* lyses its phagocytic vacuole shortly after invasion. Listeriolysin O, a potent cytolysin, accounts for the rapid lysis, as demonstrated by using *hly* mutants obtained by transposition (GAILLARD et al. 1987; PORTNOY et al. 1988; KATARIOU et al. 1987). Recently, BIELECKI et al. (1990) showed that a recombinant strain of *Bacillus subtilis* that expressed listeriolysin acquired the capacity to lyse its phagocytic vacuole within macrophages and subsequently grow intracellularly, while the wild-type strain barely survived within cells, again suggesting that lysis of the phagocytic vacuole is a prerequisite for rapid intracellular growth. These observations provide new insights into the role of hemolysins in the pathogenesis of invasive diseases. Bacteria which have the capacity to lyse their phagocytic vacuole early after entry can grow rapidly within the intracellular compartment. In contrast, bacteria that do not lyse their phagocytic vacuole have had to evolve more sophisticated means of adaptation to the harsh environment of phagosomes or phagolysosomes in order to survive and eventually grow within cells.

In addition to escaping from the phagocytic vacuole, bacteria must express factors allowing growth within the cytoplasmic compartment. Available data on such factors are limited at the moment. The capacity to synthesize folic acid via the aromatic pathway is critical since *aro* mutants grow significantly more slowly within cells (LINDBERG et al. 1990). For this reason, live attenuated strains engineered by introducing such mutations are promising as potential vaccine candidates. As will be discussed in another chapter (Maurelli et al., this volume), the porins OmpC and OmpF are also important for intracellular growth of *S. flexneri.* A deletion introduced within *ompB*, the double-component osmosensitive regulatory locus which activates transcription of porin genes (BERNARDINI et al. 1990), and transposon insertions in the structural genes *ompC* and *ompF* (M.L. BERNARDINI et al., manuscript in preparation) produce mutants that grow slowly intracellularly and have a reduced capacity to kill infected cells. The *ompB* mutant appeared negative in the Sereny test. In orally infected macaque monkeys, the *ompB* mutant elicited a mild degree of dysentery with a limited number of colonic ulcerations of smaller size than occurs in animals infected with the wild-type strain (SANSONETTI et al. 1991). However, the role of porins in the intracellular behavior of shigellae had yet to be elucidated.

On the other hand, certain other bacterial factors that might be expected to be important to intracellular growth are not. For example, no correlation has been observed between bacterial growth rate and production of Shiga or Shiga-like toxin (CLERC et al. 1987; FONTAINE et al. 1988). Neither has a correlation been found between the rate of intracellular growth of *S. flexneri* and its capacity to produce the hydroxamate siderophore aerobactin (LAWLOR et al. 1987; NASSIF et al. 1987). These data indicate either that Fe^{3+} is more available than might have been expected in the intracellular compartment or that another efficient means of iron acquisition exists in *S. flexneri*.

4 Intracellular Movement of *S. flexneri* and Cell-to-Cell Spread

The capacity to move intracellularly and spread from cell to cell appears to be another essential feature in the pathogenesis of *S. flexneri* infection of epithelial cells. After 4 h of infection, Giemsa-stained HeLa cells show numerous bacteria throughout the cell cytoplasm, not simply localized near the site of entry. This aspect of *Shigella* infection was originally demonstrated more than 20 years ago by OGAWA et al. (1968). Using phase-contrast microcinematography, they showed that intracellular shigellae undergo rapid, random movements in the cytoplasm which at times lead to the formation of protrusions. Later, MAKINO et al. (1986) demonstrated that *virG*, a plasmid locus distinct from the entry sequences, is necessary for permanent reinfection of adjacent cells in the Sereny test. More recently, it has been shown that this movement is based on the interaction between free intracellular shigellae and the host cell cytoskeleton. Treatment of infected cells by cytochalasin D, which prevents actin polymerization, blocks both intracellular movement and intercellular spread of shigellae (PAL et al. 1989).

It has been possible to demonstrate directly the interaction of the intracellular shigellae with actin by labeling infected HeLa cells with NBD-phalloidin (BERNARDINI et al. 1989). Such labeling shows that bacteria become coated with polymerized actin within 2 h of entry (Fig. 6). The bacteria appear to be located principally within the intricate network of actin cables that are situated at the level of the focal plaques of adhesions. As infection progresses, bacteria appear to lose their actin coating and to acquire a tail consisting of various lengths of polymerized actin, each approximately the diameter of a bacterium (Fig. 5). In this way, labeled actin traces the movement of the bacteria within the intracellular compartment. In some instances, bacteria appear to be located at the tip of a tail of actin, yet within a protrusion of the cell membrane that would allow passage of the bacteria from one cell to the next.

Similar intracellular activity has been described for *L. monocytogenes* (TILNEY and PORTNOY 1989; MOUNIER et al. 1990), another intracellular microorganism that has the capacity to lyse its phagocytic vacuole, move intracellularly, and spread from cell to cell. Both organisms express a surface protein that induces actin nucleation and subsequent polymerization. This process leads to the formation of a gel composed of disorganized filaments of actin. The gel then becomes localized at one pole of the bacterium, thereby forming the observed tails. It is not yet clear if the bacterium polarizes the actin gel in

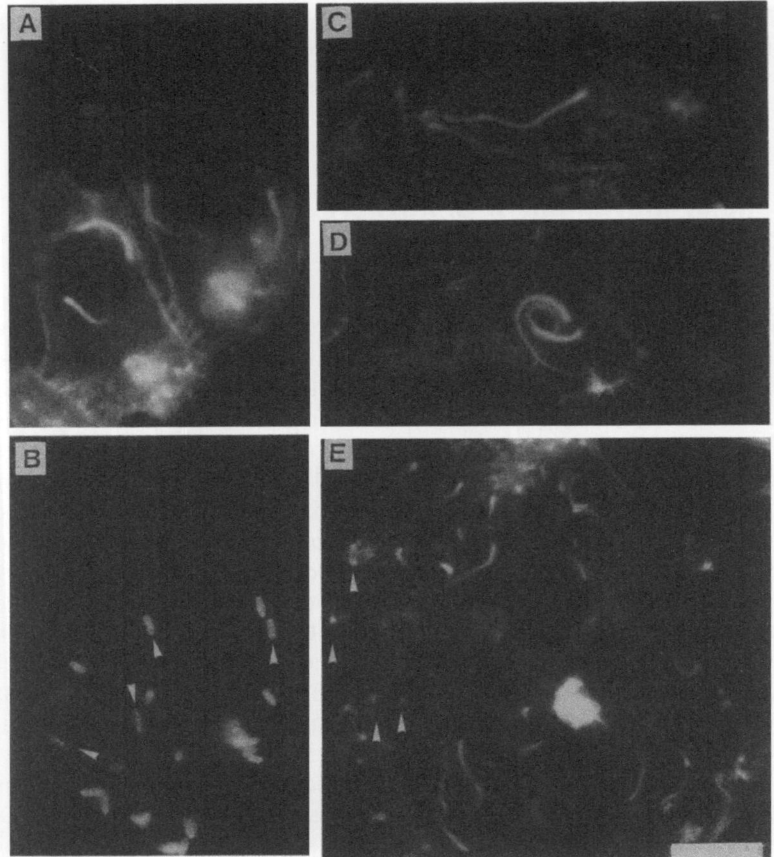

Fig. 6 A–E. HeLa cells infected by *S. flexneri*. Aspects of the Ics phenotype. **A, B** Double fluorescence labeling of infected cells showing that some bacteria (anti-lipopolysaccharide-rhodamine) indicated by arrowheads are followed by a bright tail of F-actin (NBD-phalloidin), **(A). C,D,E** Single labeling of cells with NBD-phalloidin. Intracellular shigellae coated with F-actin are indicated by arrowheads in **E.** Some aspects of intracellular movement materialized by various F-actin tails are shown in **C, D,** and **E**. *Bar*, 10 μm. Identical magnification for **A** to **D**

this way in order to propel itself forward, or if bacterial elongation leads to breakage of the gel, causing a two-phase milieu from which the bacterium is expelled, leaving behind a tail of actin. It is not known whether or not a cellular motor such as myosin (SHEETZ and SPUDICH 1983) is present within the actin network.

The capacity to spread intracellularly and from cell to cell is easily studied in vitro using the plaque assay (OAKS et al. 1985). A plaque-negative mutant of *S. flexneri* which contains a Tn*phoA* insertion in the gene *icsA* has been obtained. This mutant is unable to move intracellularly or from cell to cell. It lacks a 120-kDa outer membrane protein and does not induce actin

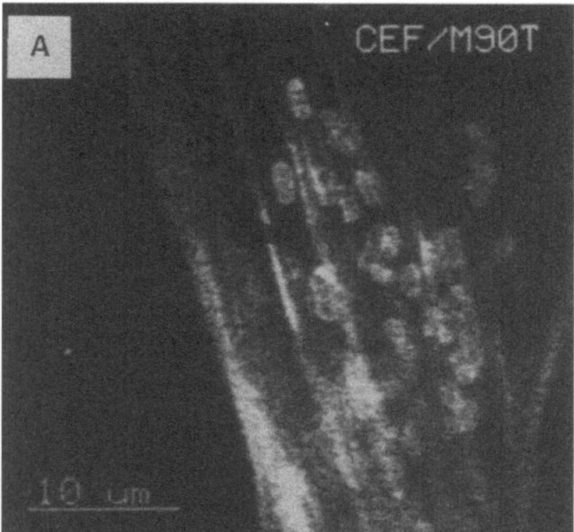

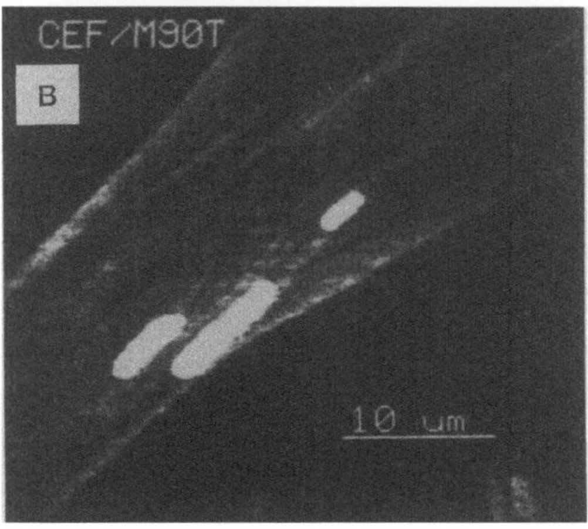

Fig. 7 A, B. Chick embryo fibroblasts infected by *S. flexneri*. Aspects of the Olm phenotype by which bacteria bind (and move) along actin stress cables. Double fluorescence labeling: NBD-phalloidin and anti-LPS-rhodamine

polymerization (BERNARDINI et al. 1989). The actin nucleation and/or polymerization activity of this protein is currently under study. *icsA* appears to be the same as *virG* (MAKINO et al. 1986). Its sequence has been published (LETT et al. 1989). It is positively regulated at the transcriptional level by the product of *kcp* (PAL et al. 1989), a gene located at 13 min on the chromosome of *S. flexneri* which was shown several years ago to be necessary for the production of a positive Sereny test by the invasive microorganism (FORMAL et al. 1971).

We have recently shown that the Ics phenotype is important in vivo. Macaque monkeys infected orally by SC560, a deletion mutant of *icsA*, developed very limited symptoms of shigellosis, as compared with the development of extensive shigellosis in animals infected with the wild-type strain. Endoscopic examination of the rectum and sigmoid colon of these animals demonstrated only a few small nodular abscesses and minor ulcerations (SANSONETTI et al. 1991). At the end of this review we will consider the implications of these observations for the understanding of the pathogenic potential of *Shigella*.

More recently, by combining microcinematography, fluorescence and confocal microscopy, and electron microscopy, we have shown (VASSELON et al. 1991) that *S. flexneri* expressed an additional type of intracellular movement that involves binding to the stress fibers of the host cell. This interaction is demonstrated in Fig. 7, which shows co-localization of shigellae on actin cables. This type of movement has been called Olm for *o*rganelle-*l*ike *m*ovement. The relevance of this movement in the pathogenesis of shigellosis and its molecular basis are not yet known.

5 Killing of Host Cells by *Shigella*

The mechanisms by which shigellae destroy the host cell are still poorly understood. Investigators had previously assumed that Shiga toxin accounted for host cell killing (HALE and FORMAL 1981). Shiga toxin is a potent cytotoxin which blocks protein biosynthesis by destroying ribosomes via *N*-glycosylation of RNA, a mechanism also employed by the toxins ricin and modeccin (ENDO and TOURUGI 1987). These activities of Shiga toxin are extensively discussed by O'BRIEN et al. (this volume). However, much recent evidence does not support the hypothesis of a major role for this toxin in the rapid killing of invaded cells. In vitro, isolates of *S. flexneri* are able to kill eukaryotic cells very efficiently, even if they do not produce significant amounts of this toxin (CLERC et al. 1987). Using an invasion assay with J774 macrophages as targets, rapid and efficient killing of cells infected with *S. flexneri* correlated with the expression of the invasive phenotype. Macrophages were protected by cytochalasin D, demonstrating that expression of this phenotype required bacteria to be intracellular (CLERC et al. 1987). In a similar invasion assay, we were unable to demonstrate a significant difference in the rate of macrophage killing by a strain of *S. dysenteriae* 1 and its Tox$^-$ mutant (FONTAINE et al. 1988). Moreover, in vitro, the killing process induced by bacterial invasion is much faster (occurring 1–2 h after entry of the bacteria) than the cytotoxic process caused by Shiga toxin added extracellularly (occurring at least 8 h after entry).

Metabolic alterations observed during the early stages of cell killing by invasive *S. flexneri* include a rapid decrease in the intracellular concentration of adenosine triphosphate (ATP, occurring within 30 min after entry), as well as an increase in pyruvate and a decrease in lactate concentrations, suggesting the discontinuation of cell respiration and fermentation, respectively (SANSONETTI and MOUNIER 1987). The molecular basis of these observations is as yet unknown. TEM performed on J774 macrophages infected by *S. flexneri* has shown that early after entry invasive bacteria lysed their phagocytic vacuoles and localized close to mitochondria, which appeared condensed with destruction of their internal compartments. Mitochondria may therefore be a major intracellular target for the pathogen.

However, Shiga toxin has additional modes of action in vivo. Based on experiments performed in macaque monkeys infected orally by either a Tox^+ isolate of *S. dysenteriae* 1 or a Tox^- mutant derived from the same isolate, it has been possible to demonstrate that secretion of Shiga toxin within infected tissues caused severe alterations of capillaries of the lamina propria of colonic villi (FONTAINE et al. 1988). The toxin may therefore cause ischemic and hemorrhagic colitis in addition to the invasive process. This may, largely, explain the greater severity of bacillary dysentery caused by *S. dysenteriae* 1.

6 Conclusion: From In Vitro to In Vivo

Many links are still missing in our understanding of the pathogenesis of shigellosis. The ability of the bacteria to enter host cells is essential to virulence. Strains that have lost the expression of this entry phenotype are avirulent even in the most definitive virulence assays. At present a major question is where in the human intestine does entry take place? It is not clear why invasion does not occur at the level of the small intestine. It is also not clear, at the colonic level, whether or not bacteria penetrate at the apical pole of enterocytes. We have mentioned previously that bacteria bind to and invade basolateral pole of enterocyte-like Caco-2 cells, thus suggesting that the bacterial receptor is not expressed at the surface of the brush border. How do shigellae gain access to the laterobasal pole of enterocytes under these circumstances? One possibility is that they invade colonic crypts where enterocytes are less differentiated. In addition, when macaque monkeys are infected by an *icsA* mutant of *S. flexneri*, they develop only very limited symptoms of dysentery (SANSONETTI et al. 1991). Colonoscopic examination reveals the presence of a limited number of small nodular abscesses or tiny ulcerations. Biopsy of these lesions demonstrates that they almost always overlay lymphoid follicles, thus suggesting that entry occurs via the colonic equivalent of Peyer's patches. A mutant strain that has lost its capacity to spread intracellularly and from cell to cell is severely impaired in its virulence, thus indicating that this phenotype is essential in vivo as well as in vitro. In addition, this mutant points to the site of entry of shigellae within the epithelium. This site appears to correspond to the colonic equivalent of Peyer's patches, localized areas of the colonic mucosa rich in M cells (BYE et al. 1984). These areas may represent the "Trojan horse" of the colon, allowing entry of invasive microorganisms, into the epithelium. As summarized in Fig. 8, following penetration into the epithelium via M cells, bacteria may then spread from one cell to another by expressing their Ics

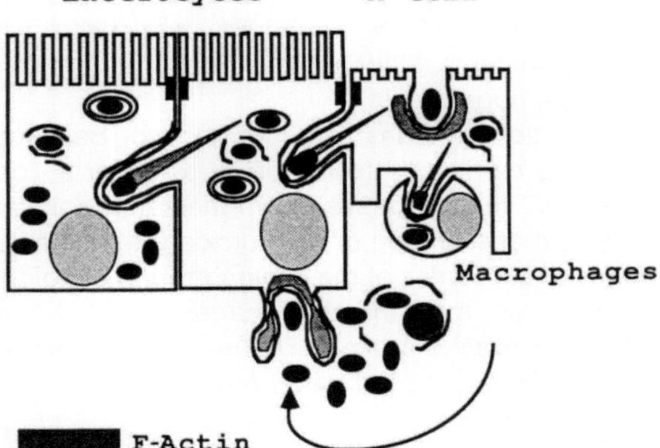

Fig. 8. Provisional scheme of colonic epithelial colonization by *S. flexneri*

phenotype. Subepithelial spread to resident macrophages within the lamina propria may also allow, after lysis of these cells, retrograde reinfection of enterocytes by their basolateral pole. This scheme is still hypothetical; however, recent data obtained in the rabbit ligated ileal loop model indicate that M cells represent the initial site of entry of shigellae (WASSEF et al. 1989).

Acknowledgements. I wish to thank all my colleagues for help and advice in the preparation of this paper, Françoise Doher for typing the manuscript, and Marcia Goldberg for careful reading and helpful suggestions.

References

Adler B, Sasakawa C, Tobe T, Makino S, Komatsu K, Yoshikawa M (1989) A dual transcriptional activation system for the 230kb plasmid genes coding for virulence associated antigens of *Shigella flexneri*. Mol Microbiol 3: 627–635

Barak LS, Yocum RR, Nothnagel EA, Webb WW (1980) Fluorescence staining of the actin cytoskeleton in living cells with 7-nitrobenz-2-oxa-1, 3-diazole-phallacidin. Proc Natl Acad Sci USA 77: 980–984

Baudry B, Maurelli AT, Clerc P, Sadoff JC, Sansonetti PJ (1987) Localization of plasmid loci necessary for the entry of *Shigella flexneri* into HeLa cells, and characterization of one locus encoding four immunogenic polypeptides. J Gen Microbiol 133: 3403–3413

Baudry B, Kaczorek M, Sansonetti PJ (1988) Nucleotide sequence of the invasion plasmid antigen B and C genes (*ipa*B and *ipa*C) of *Shigella flexneri*. Microb Pathog 4: 345–357

Bernardini ML, Mounier J, d'Hauteville H, Coquis-Rondon M, Sansonetti PJ (1989) Identification of *ics*A, a plasmid locus of *Shigella flexneri* that governs intra- and intercellular spread through interaction with F-actin. Proc Natl Acad Sci USA 86: 3867–3871

Bernardini ML, Fontaine A, Sansonetti PJ (1990) The two-component regulatory system *OmpR-EnvZ* controls the virulence of *Shigella flexneri*. J Bacteriol 172: 6274–6281

Bielecki J, Youngman P, Connelly P, Portnoy DA (1990) *Bacillus subtilis* expressing a hemolysin gene from *Listeriá monocytogenes* can grow in mammalian cells. Nature 345: 175–176

Buysse JM, Stover CK, Oaks EV, Venkatesan M, Kopecko DJ (1987) Molecular cloning of invasion plasmid antigen (ipa) genes from *Shigella flexneri*: analysis of ipa gene products and genetic mapping. J Bacteriol 169: 2561–2569

Buysse JM, Venkatesan JA, Mills JA, Oaks EV (1990) Molecular characterization of a *trans*-acting positive effector (*ipaR*) of invasion plasmid antigen synthesis in *Shigella flexneri* serotype 5. Microb Pathog 8: 197–211

Bye WA, Allan CH, Trier JS (1984) Structure, distribution and origin of M cells in Peyer's patches of mouse ileum. Gastroenterology 86: 789–801

Clerc P, Sansonetti PJ (1987) Entry of *Shigella flexneri* into HeLa cells: evidence for direct phagocytosis involving actin polymerization and myosin accumulation. Infect Immun 55: 2681–2688

Clerc P, Sansonetti PJ (1989) Evidence for clathrin mobilization during directed phagocytosis of *Shigella flexneri* by HEp2 cells. Microb Pathog 7: 329–336

Clerc P, Baudry B, Sansonetti PJ (1986) Plasmid-mediated contact hemolytic activity in *Shigella* species: correlation with penetration into HeLa cells. Ann Inst Pasteur Microbiol 137A(3): 267–278

Clerc P, Ryter A, Mounier J, Sansonetti PJ (1987) Plasmid mediated early killing of eucaryotic cells by *Shigella flexneri* as studied by infection of J774 macrophages. Infect Immun 55: 521–527

Endo Y, Tourugi K (1987) RNA *N*-glycosidase activity of ricin A-chain. J Biol Chem 262: 8128–8131

Fontaine A, Arondel J, Sansonetti PJ (1988) Role of Shiga-toxin in the pathogenesis of shigellosis as studied using a Tox^- mutant of *Shigella dysenteriae* 1. Infect Immun 56: 3099–3109

Formal SB, Gemski P, Baron LS, LaBrec EH (1971) A chromosomal locus which controls the ability of *Shigella flexneri* to evoke keratoconjunctivitis. Infect Immun 3: 73–79

Gaillard JL, Berche P, Mounier J, Richard S, Sansonetti PJ (1987) In vitro model of penetration and intracellular growth of *Listeria monocytogenes* in the human enterocyte-like cell line CaCo2. Infect Immun 55: 2822–2829

Gots RE, Formal SB, Giannella RA (1974) Indomethacin inhibition of Salmonella typhimurium, *Shigella flexneri*, and cholera-mediated rabbit ileal secretion. J Infect Dis 130: 280–284

Hale TL (1986) Invasion of epithelial cells by Shigella. In microbial invasion of non-phagocytic cells. Ann Inst Pasteur Microbiol 137A: 311–314

Hale TL, Bonventre PF (1979) Shigella infection of Henle Intestinal epithelial cells: role of the bacteria. Infect Immun 24: 879–886

Hale TL, Formal SB (1981) Protein synthesis in HeLa or Henle 407 cells infected with *Shigella dysenteriae* 1, *Shigella flexneri* 2a, or *Salmonella typhimurium* W118. Infect Immun 46: 470–475

Hale TL, Morris RE, Bonventre PF (1979) Shigella infection of Henle intestinal epithelial cells: role of the host cell. Infect Immun 24: 887–894

Hale TL, Oaks EV, Formal SB (1985) Identification and antigenic characterization of virulence-associated, plasmid-coded proteins of Shigella spp. and enteroinvasive *Escherichia coli*. Infect Immun 50: 620–629

Hromockyj AE, Maurelli AT (1989) Identification of Shigella invasion genes by construction of temperature-regulated *inv*::*lacZ* operon fusions. Infect Immun 57: 2963–2970

Isberg RR, Leong JM (1990) Multiple beta-1 chain integrins are receptors for Invasin, a protein that promotes bacterial penetration into mammalian cells. Cell 60: 861–871

Katariou Ş, Metz P, Hof H, Goebel W (1987) Tn916-induced mutations in the hemolysin determinant affecting virulence of *Listeria monocytogenes*. J Bacteriol 169: 1291–1297

Koster F, Levin J, Walker L, Tung KSK, Gilman RH, Rahaman MM, Majid MA, Islam S, Williams RC (1978) Hemolytic-uremic syndrome after shigellosis. Relation to endotoxemia and circulating immune complexes. N Engl J Med 298: 927–933

LaBrec EH, Schneider H, Magnani TJ, Formal SB (1964) Epithelial cell penetration as an essential step in pathogenesis of bacillary dysentery. J Bacteriol 88: 1503–1518

Lawlor KM, Daskaleros PA, Robinson RE, Payne SM (1987) Virulence of iron transport mutants of *Shigella flexneri* and utilization of host iron compounds. Infect Immun 55: 594–599

Lett MC, Sasakawa C, Okada N, Sakai T, Makino S, Yamada M, Komatsu K, Yoshikawa M (1989) *vir*G, a plasmid coded virulence gene of *Shigella flexneri*: identification of the VirG protein and determination of the complete coding sequence. J Bacteriol 171: 353–359

Lindberg AA, Karnell A, Pal T, Sweiha H, Hultenby K, Stocker BAD (1990) Construction of an auxotrophic *Shigella flexneri* strain for use as a live vaccine. Microb Pathog 8: 433–440

Louvard D (1989) The function of the major cytoskeletal components of the brush border. Curr Opin Cell Biol 1: 51–57

Makino S, Sasakawa C, Kamata K, Kurata T, Yoshikawa M (1986) A genetic determinant required for continuous reinfection of adjacent cells on large plasmid in *Shigella flexneri* 2a. Cell 46: 551–555

Maurelli AT, Baudry B, d'Hauteville H, Sansonetti PJ (1985) Cloning of plasmid DNA sequences involved in invasion of HeLa cells by *Shigella flexneri*. Infect Immun 49: 164–171

Mooseker MS, Coleman TR (1989) The 110-kD protein-calmodulin complex of the intestinal microvillus (brush border myosin I) is a mechanoenzyme. J Cell Biol 108: 2395–2400

Mounier J, Ryter A, Coquis-Rondon M, Sansonetti PJ (1990) Intracellular and cell to cell spread of *Listeria monocytogenes* involves interaction with F-actin in the enterocyte-like cell line Caco-2. Infect Immun 58: 1048–1058

Nassif X, Mazert MC, Mounier J, Sansonetti PJ (1987) Evaluation with an iuc::Tn10 mutant of the role of aerobactin production in the virulence of *Shigella flexneri*. Infect Immun 55: 1963–1967

Neill RJ, Gemski P, Formal SB, Newland JW (1988) Deletion of the shiga toxin gene in a chlorate-resistant derivative of *Shigella dysenteriae* type 1 that retains virulence. J Infect Dis 158: 737–741

Oaks EV, Wingfield ME, Formal SB (1985) Plaque formation by *Shigella flexneri*. Infect Immun 48: 124–129

Oaks EV, Hale TL, Formal SB (1986) Serum immune response to Shigella protein antigens in rhesus monkeys and humans infected with *Shigella* spp. Infect Immun 53: 57–63

O'Brien AD, Holmes RK (1987) Shiga and Shiga-like toxins. Microbiol Rev 51: 206–220

Ogawa H, Nakamura A, Nakaya R (1968) Cinematographic studies of tissue cell cultures infected with *Shigella flexneri* . Jpn J Med Sci Biol 21: 259–273

Pal T, Newland JW, Tall D, Formal SB (1989) Intracellular spread of *Shigella flexneri* associated with the kcpA locus and a 140-kilodalton protein. Infect Immun 57: 477–486

Portnoy DA, Jacks PS, Hinrichs DJ (1988) Role of hemolysin for the intracellular growth of *Listeria monocytogenes*. J Exp Med 167: 1459–1471

Prentki P, Kirsch MM (1984) In vitro insertional mutagenesis with a selectable DNA fragment. Gene 29: 303–313

Relman D, Tuomanen E, Falkow S, Golenbock DT, Saukkonen K, Wright SD (1990) Recognition of a bacterial adhesin by an integrin: macrophage CR3 alphaM/beta2, CD11b/CD18, binds filamentous hemagglutinin of *Bordetella pertussis*. Cell 61: 1375–1382

Rousset M (1986) The human colon carcinoma cell lines HT29 and Caco-2: two in vitro models for the study of intestinal differenciation. Biochimie 68: 1035–1040

Ruoslahti E, Pierschbacher MD (1987) New perspectives in cell adhesion: RGD and integrins. Science 238: 491–497

Sansonetti PJ, Mounier J (1987) Metabolic events mediating early killing of host cells infected by *Shigella flexneri*. Microb Pathog 3: 53–61

Sansonetti PJ, Kopecko DJ, Formal SB (1982) Involvement of a plasmid in the invasive ability of *Shigella flexneri*. Infect Immun 35: 852–860

Sansonetti PJ, Hale TL, Dammin GI, Kapper C, Collins HH, Formal SB (1983) Alterations in the pathogenesis of *Escherichia coli* K12 after transfer of plasmid and chromosomal genes from *Shigella flexneri*. Infect Immun 39: 1392–1402

Sansonetti PJ, Ryter A, Clerc P, Maurelli AT, Mounier J (1986): multiplication of *Shigella flexneri* within HeLa cells: lysis of the phagocytic vacuole and plasmid mediated contact hemolysis. Infect Immun 51: 461–465

Sansonetti PJ, Arondel J, Fontaine A, d'Hauteville H, Bernardini L (1991) ompB (osmo-regulation) and icsA (cell to cell spread) mutants of *Shigella flexneri*: vaccine candidates and probes to study the pathogenesis of shigellosis. Vaccine (in press)

Sakai T, Sasakawa, Makino S, Yoshikawa M (1986) DNA sequence and product analysis of the vir F locus responsible for Congo red binding and cell invasion in *Shigella flexneri*. Infect Immun 54: 395–402

Sasakawa C, Kamata K, Sakai T, Makino S, Yamada M, Okada N, Yoshikawa M (1988) Virulence-associated genetic regions comprising 31 kilobases of the 230-kilobase plasmid in *Shigella flexneri* 2a. J Bacteriol 170: 2480–2484

Sasakawa C, Adler B, Tobe T, Okada N, Nagai S, Komatsu K, Yoshikawa M (1989) Functional organization and nucleotide sequence of virulence region-2 on the large virulence plasmid of *Shigella flexneri* 2a. Mol Microbiol 170: 2480–2484

Sereny B (1957) Experimental keratoconjunctivitis shigellosa. Acta Microbiol Acad Sci Hung 4: 367–376

Sheetz MP, Spudich JA (1983) Movement of myosin-coated fluorescent beads on actin cables in vitro. Nature 303: 31–35

Sheterline P, Rickard JE, Richards RC (1984) Fc receptor directed phagocytic stimuli induce transient actin assembly at an early stage of phagocytosis in neutrophil leucocytes. Eur J Cell Biol 34: 80–87

Small PLC, Isberg RR, Falkow S (1987) Comparison of the ability of enteroinvasive *Escherichia coli*, *Yersinia pseudotuberculosis* and *Yersinia enterocolitica* to enter and replicate within Hep-2 cells. Infect Immun 55: 1674–1679

Stendahl OI, Hartwig JH, Brotschi EA, Stossel TP (1980) Distribution of actin-binding protein and myosin in macrophages during spreading and phagocytosis. J Cell Biol 84: 215–224

Tanenbaum SW (ed) (1978) Cytochalasins: biochemical and cell biological aspects. North-Holland, Amsterdam

Takeuchi A, Spring H, Labrec EH, Formal SB (1965) Experimental bacillary dysentery: an electron microscopic study of the response of intestinal mucosa to bacterial invasion. Am J Pathol 4: 1011–1044

Takeuchi A, Formal SB, Sprinz H (1968) Experimental acute colitis in the rhesus monkey following peroral infection with *Shigella flexneri*. Am J Pathol 52: 503–529

Tilney LG, Portnoy DA (1989) Actin filaments and the growth, movement, and spread of the intracellular bacterial parasite, *Listeria monocytogenes*. J Cell Biol 109: 1597–1608

Vasselon T, Mounier J, Prevost MC, Hellio R, Sansonetti PJ (1991) A stress fiber-based movement of *Shigella flexneri* within cells. Infect Immun (in press)

Wassef JS, Keren DF, Mailloux JL (1989) Role of M cells in initial bacterial uptake and in ulcer formation in the rabbit intestinal loop model in shigellosis. Infect Immun 57: 858–863

Watanabe H, Arakawa E, Ito K, Kato JI, Nakamura A (1990) Genetic analysis of an invasion region by use of a Tn3-lac transposon and identification of a second positive regulator gene, inv E, for cell invasion of *Shigella sonnei*: significant homology of inv E with par B of plasmid PI. J Bacteriol 172: 619–629

The Large Virulence Plasmid of *Shigella*

C. SASAKAWA[1], J. M. BUYSSE[2], and H. WATANABE[3]

1 Introduction

Invasion of human colonic epithelial cells constitutes one of the earliest steps in the pathogenesis of dysentery caused by *Shigella* species and enteroinvasive strains of *Escherichia coli* (EIEC). Following the invasion of target cells by virulent bacteria, their subsequent multiplication, intracellular movement, and intercellular spread result in a focus of *Shigella* infection that is characterized by severe desquamation and ulceration of the mucosa. The first indication that invasion of colonic epithelial cells was critical in the development of the dysenteric syndrome was presented by LABREC et al. (1964) who showed that a virulent, translucent colony morphology strain of *Shigella flexneri* 2a (2457T)

[1] Department of Bacteriology, Institute of Medical Science, University of Tokyo, Minato-ku, Tokyo 108, Japan
[2] Department of Bacterial Immunology, Division of Communicable Diseases and Immunology, Walter Reed Army Institute of Research, Walter Reed Army Medical Center, Washington, DC 20307–5100, USA
[3] Department of Bacteriology, National Institute of Health, Shinagawa-ku, Tokyo 141, Japan

Current Topics in Microbiology and Immunology, Vol. 180

was able to penetrate colonic epithelia in both rhesus monkeys and opiated guinea pigs, whereas an avirulent opaque variant (2457O) did not. However, it was not until the early 1980s and the pioneering work of SANSONETTI and colleagues that the essential role of a large (100–140 MDa) plasmid, found in all virulent strains of *Shigella* and EIEC, was established for the invasion phenotype (KOPECKO et al. 1980; SANSONETTI et al. 1981, 1982, 1983b; HARRIS et al. 1982). Subsequent rapid progress in the study of virulence-related (*vir*) genes carried by this large invasion plasmid has been complemented by the development of recombinant DNA technology, in vitro assay systems for the various virulence phenotypes, and advances in cellular and molecular biology. As will be reviewed in this chapter, our knowledge of the role of the large plasmid in determining the virulence properties of *Shigella* has accumulated substantially over the last few years; however, a full characterization of the *vir* genes has not yet been realized. Studies to date have indicated that the invasion plasmid encodes genes for: (a) ligands that are involved in the adherence of bacteria onto the surface of target epithelial cells (PAL and HALE 1989); (b) the production of invasion plasmid antigens (Ipa) that have a direct role in the *Shigella* invasion process (BUYSSE et al. 1987; BAUDRY et al. 1987; SASAKAWA et al. 1989); (c) transport or processing functions that ensure the correct surface expression of the Ipa proteins (HROMOCKYJ and MAURELLI 1989); (d) the induction of endocytic uptake of bacteria (CLERC et al. 1987) and disruption of endocytic vacuoles (SANSONETTI et al. 1986); (e) the intracellular and intercellular spreading phenotype (MAKINO et al. 1986; BERNARDINI et al. 1989); and (f) the regulation of plasmid-encoded *vir* genes (SAKAI et al. 1988; ADLER et al. 1989; BUYSSE et al. 1990; WATANABE et al. 1990).

2 Structure and Function of *vir* Genes on the Large Plasmid

2.1 Overview of Early Molecular Studies

The first reports of plasmid-linked virulence properties were published by SANSONETTI and coworkers in the early 1980s. After demonstrating that form I antigen expression and invasive ability could be restored to *S. sonnei* form II cells via conjugal mobilization of the 180-kb *S. sonnei* plasmid, several groups went on to demonstrate that all virulent shigellae and EIEC strains harbor a large plasmid that mediates epithelial cell invasion (reviewed by KOPECKO et al. 1985). Initial molecular analysis of these invasion plasmids consisted of detailed comparisons of restriction enzyme fragment profiles, DNA homology studies and characterization of unique plasmid-encoded proteins. Although restriction enzyme patterns of the plasmid varied among species and serotypes, the molecules were found to have extensive DNA homology as determined by

Southern hybridization analysis (HALE et al. 1983; SANSONETTI et al. 1983a). Transfer of a wild-type invasion plasmid into avirulent, plasmid-cured or plasmid-deleted *Shigella* strains and into *E. coli* K12 established that the invasive phenotype and its genetic determinants were plasmid encoded (SANSONETTI et al. 1983b; MAURELLI et al. 1984a; WATANABE and NAKAMURA 1985; SASAKAWA et al. 1986a).

To identify DNA sequences on the large plasmid responsible for the virulence phenotype of *Shigella*, two different approaches have been exploited, namely cosmid cloning and insertion mutagenesis. MAURELLI et al. (1985) used a multiple-copy cosmid vector, pJB8 (ISH-HORWICZ and BURKE 1981), to clone genes from the invasion plasmid (pWR100) of *S. flexneri* 5, required for penetration of cultured HeLa cells. A resulting 44-kb cosmid clone, designated pHS4108 (Fig. 1), restored the invasion phenotype to a mutant of *S. flexneri* 5 lacking the large plasmid. DNA hybridization studies established that the 44-kb

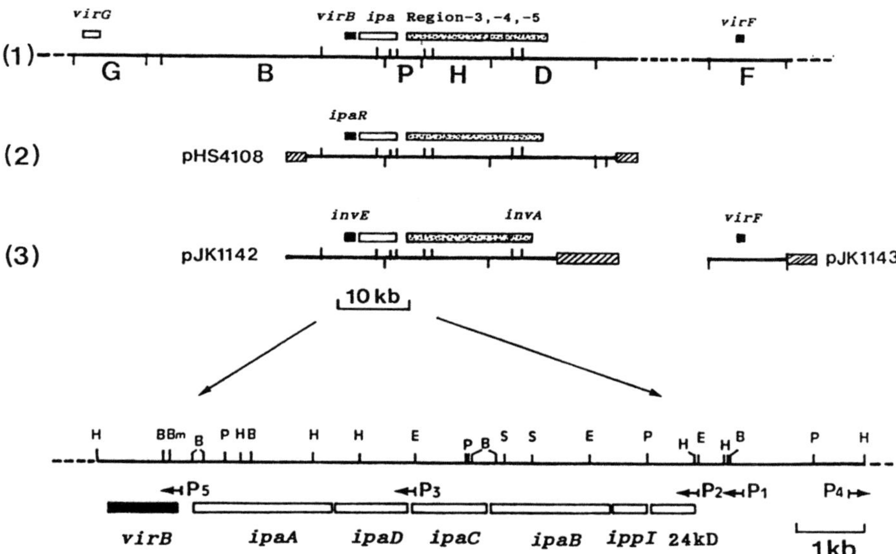

Fig. 1. The genetic organization of the virulence regions on the large plasmid of *S. flexneri*. (*1*), The seven virulence regions defined by Tn5 insertions into pMYSH6000 of *S. flexneri* 2a YSH6000 (SASAKAWA et al. 1986a); (*2*), pSH4108, a cosmid clone of pWR100 essential for the Inv$^+$ phenotype for *S. flexneri* 5 M90T (MAURELLI et al. 1985); (*3*), pJK1142, a cosmid clone of pSS120 essential for the Inv$^+$ phenotype for *S. sonnei* HW383 (WATANABE and NAKAMURA 1985) and pJK1143, the cloned *virF* region (KATO et al. 1989). The *vertical bars over* and *beneath* the *thick lines* for (*1*)–(*3*) indicate the *Eco*RI and *Sal*I sites, respectively. The *bars* for *Eco*RI sites on (*1*) are shown only within the region corresponding to pHS4108 and pJK1142. The *dotted boxes* over the *thick lines* of pHS4108 and pJK1142 indicate the invasion-associated genetic regions corresponding to regions 3, 4, and 5 on pMYSH6000. The *hatched boxes* of pHS4108, pJK1142, and pJK1143 represent the vectors. The *bottom line* shows the restriction map of the *ipa* operon and the *virB* (*virR* and *invE*) regions. P_1–P_5 indicate the promoter regions. Restriction sites are: *B, Bgl*II; *Bm, Bam*HI; *E, Eco*RI; *H, Hin*dIII; *P, Pst*I; *S, Sal*I

segment is highly conserved among other large plasmids of *Shigella* and EIEC. However, transformants carrying pHS4108 lacked the ability to spread intercellularly, suggesting that pWR100 encoded additional *vir* gene(s) associated with the intercellular spreading phenotype. pHS4108 synthesized four antigens, recognized by convalescent monkey antisera, of 70 kDa, 62 kDa, 41 kDa, and 37 kDa [subsequently designated IpaA, IpaB, IpaC, and IpaD, respectively, by BUYSSE et al. (1987)]; and the expression of the antigens was found to be temperature regulated as observed in the wild-type M90T parent (MAURELLI et al. 1984b). Tn5 insertion analysis of the 44-kb clone suggested that the corresponding *ipaA, ipaB, ipaC,* and *ipaD* genes comprise one or more operons, and that three of the genes (*ipaB, ipaC,* and *ipaD*) are required for the invasion phenotype (BAUDRY et al. 1987). A more detailed genetic organization of the *ipaB, ipaC,* and *ipaD* genes was determined by BUYSSE et al. (1987), and the resulting restriction map resembles that determined for region 2 of pMYSH6000 in *S. flexneri* 2a (SASAKAWA et al. 1986a, 1988) (Fig. 1; see Sect. 2.2).

The invasion plasmid of *S. flexneri* 2a (pMYSH6000) was initially used to make a *SalI* restriction map (SASAKAWA et al. 1986a). The analysis of contiguous *SalI* fragments in spontaneous deletion mutants of pMYSH6000 or in various partial *SalI* digests of pMYSH6000 made it possible to order the 23 *SalI* fragments, designated A–T. By using a thermosensitive replication mutant of R388 carrying Tn*5* (SASAKAWA and YOSHIKAWA 1987), SASAKAWA et al. (1986b) obtained over 300 Tn*5* insertions in pMYSH6000, and each Tn*5* insertion was assigned to one of the 23 *SalI* fragments; only *SalI* fragments B, D, F, G, H, and P contained Vir$^-$ Tn*5* insertions. Assaying various virulence-associated phenotypes, including the Sereny test for the ability to provoke keratoconjunctivitis in the eyes of guinea pigs (SERENY 1957) or in those of mice (Vir$^+$) (MURAYAMA et al. 1986), invasion in a LLC-MK2 cell model (Inv$^+$) (WATANABE and NAKAMURA 1985), and the ability to bind Congo red (Pcr$^+$) (MAURELLI et al. 1984a), the collection of Tn*5* mutants was divided into three classes:(I) Vir$^-$ Inv$^-$ Pcr$^-$; (II) Vir$^-$ Inv$^-$ Pcr$^+$; and (III) Vir$^-$ Inv$^+$ Pcr$^+$. Five virulence-associated (*vir*) regions, tentatively named regions 1–5, comprising a 31-kb DNA segment of pMYSH6000 were located on the contiguous *SalI* fragments B–P–H–D. Insertion mutants in these regions belong to class I, except those in region 2 (containing the *ipa* genes; see Sect. 2.2) which belong to class II. A small 1.0-kb *vir* region, identified in *SalI* fragment F, was designated *virF* (see the Sect. *virF*), and insertion mutants in this locus belong to class I. Class III was identified by insertions into *SalI* fragment G (containing the *virG* or *icsA* locus; see Sect. 2.6), which is located 20 kb from the 3′ end of region 1; the Vir$^-$ Inv$^+$ Pcr$^+$ phenotype of class III mutations indicates that *virG* is not involved in invasion. All of the *vir* regions identified in pMYSH6000 are highly conserved among other large plasmids, and the physical structure of the 31-kb virulence DNA segment of pMYSH6000 resembles that of pHS4108 (Fig. 1) (MAURELLI et al. 1985; SASAKAWA et al. 1988).

Corresponding *vir* genes of the *S. sonnei* invasion plasmid pSS120 were initially examined by Tn*1* insertion mutagenesis (WATANABE and NAKAMURA 1986).

By introducing various *Hin*dIII fragments of the large plasmid into each of the Tn*1* insertion mutants, contiguous 2.6- and 4.1-kb *Hin*dIII fragments were found to restore the Inv$^+$ phenotype. The 4.1-kb *Hin*dIII fragment encodes a 38 kDa protein which was designated InvA. The restriction map of the *invA*-containing fragments is similar to that of region-5 of pMYSH6000. A 37-kb DNA section of pSS120 required for bacterial invasion has recently been cloned into a F plasmid-derived cosmid vector, pJK292 (KATO et al. 1989). Unlike pHS4108 (MAURELLI et al. 1985), this 37-kb low-copy cosmid clone, designated pJK1142 (Fig. 1), does not confer the Inv$^+$ phenotype on strains of *S. sonnei* lacking the invasion plasmid. Only in the presence of the *virF* gene (KATO et al. 1989) does the S. sonnei strain carrying pJK1142 become invasive, indicating that the 37-kb cosmid clone of pSS120 lacks the *virF* positive regulatory gene (SAKAI et al. 1988; see Sect. 2.3). Thus it is likely that the reason for the discrepancy in the restoration of the Inv$^+$ phenotype between the 44-kb cosmid clone of pWR100 (pHS4108) and the 37-kb cosmid clone of pSS120 (pJK1142) rests on the difference in copy number of the cosmid vectors used rather than on differences in the genes carried. Though the strategies used to identify *vir* genes from the three large plasmids were different, the results obtained from each plasmid were complementary.

2.2 Invasion Plasmid Antigen Genes

After it was established that the large virulence plasmid encodes the essential genetic determinants for epithelial cell invasion, HALE et al. (1983) identified specific invasion plasmid-encoded proteins in anucleate *S. flexneri* minicells, carrying the plasmid, that were capable of penetrating cultured HeLa cells. Analysis of these proteins revealed a complement of seven polypeptides, designated a–g, which were unique to the invasion plasmid of *S. flexneri* serotype 5. Subsequent Western blot analysis, using monkey or human convalescent sera specifically absorbed to retain palsmid antigen specificity, demonstrated that polypeptides a, b, c, and d were antigens that were immunologically similar among the various strains of *Shigella* and EIEC. Antigens a–d, and an additional plasmid-encoded 130-kDa protein, are the principal immunogens detected by serum of acute and convalescent shigellosis patients (HALE et al. 1985; OAKS et al. 1986), and an accumulating body of evidence supports the idea that these invasion plasmid antigens play a crucial role in establishing the bacteria within host colonic epithelial cells.

As described in Sect. 2.1, saturation Tn*5* mutagenesis has been used to define five invasion-associated genetic regions on the *S. flexneri* 2a invasion plasmid. The restriction enzyme map of these contiguous, linked regions corresponds with the maps determined for the Inv$^+$ cosmids derived from *S. flexneri* 5 pWR100 and *S. sonnei* pSS120 (see Fig. 1). In view of the demonstrated conservation of these sequences detected by Southern blot analysis of

Shigella species and EIEC invasion plasmid DNA (VENKATESAN et al. 1988b; SASAKAWA et al. 1988), it follows that regions 1–5 define essential genetic determinants for invasion carried by the previously isolated Inv^+ cosmids. Tn*5* mutagenesis of one of these cosmids (pHS4108), coupled with Western blot and HeLa cell invasion analysis of the mutants, revealed that insertions which caused a reduction in the expression of polypeptides b, c, and d significantly decreased the invasion ability of the organisms (MAURELLI et al. 1985; BAUDRY et al. 1987). Insertions that abolished the expression of polypeptide a, however, had no effect on the invasive phenotype, suggesting that protein a is not an invasion determinant. In a recent study by SASAKAWA et al. (1989) plasmid constructs encoding various combinations of the b, c, and d polypeptides were used to complement Tn5 mutations in region 2; thses studies confirmed that the three antigens are essential components of the Inv^+ phenotype.

In 1987, BUYSSE et al., established the genetic organization of the a–d immunogens and designated the corresponding alleles as *i*nvasion *p*lasmid *a*ntigen (*ipa*) genes *ipaA, ipaB, ipaC,* and *ipaD*, respectively. The λgt11 cloning vector was used to create an expression library of the *S. flexneri* 5 pWR110 invasion plasmid, and recombinant phages (λgt11Sf1) producing pWR110-encoded polypeptide antigens were identified with rabbit antisera specific for the IpaB, IpaC, and IpaD immunogens. Recombinants encoding the synthesis of complete, truncated, and β-galactosidase fusion versions of each of the three antigens were isolated, and insert DNA purified from the recombinants was used in Southern blot analysis of pWR100 to establish the gene order and restriction map for these loci. This map, and subsequent refinements (BAUDRY et al. 1987; BUYSSE et al. 1990), revealed that four *ipa* genes are clustered in a 6-kb segment of the invasion plasmid in the transcriptional order *ipaB–ipaC–ipaD–ipaA*. Affinity-purified antibodies, prepared from truncated peptides of the IpaB and IpaC antigens, were used to show that the Ipa proteins were immunologically distinct, but that each molecule contained separable epitope units. The versatility of combining the λgt11 expression system and specific immune sera to dissect invasion plasmid antigen genes was further demonstrated by the characterization of a series of λgt11 clones that encoded a previously unrecognized antigen, IpaH (see Sect. 2.7).

The λgt11 cloning system allows the construction and expression of recombinants encoding portions of a protein as small as a single epitope (STOVER et al. 1987). MILLS et al. (1988) created a λgt11 epitope library of the IpaB and IpaC proteins and were able to genetically map each epitope defined by a panel of IpaB- and IpaC-specific monoclonal antibodies (MAbs). Three contiguous IpaB epitopes (2F1, 1H4, and 4C8) were located on a 28-kDa peptide encompassing a 700-bp amino terminal segment of the *ipaB* gene. Similarly, a 16-kDa amino terminal peptide of IpaC contained three clustered epitopes (5H1, 5B1, and 9B6) in a 640-bp segment of the gene that were separated by 50 bp from a single epitope near the middle of IpaC, designated 2G2. Interestingly, the genetic map position of these epitopes exactly overlapped regions of strong hydrophilicity predicted by DNA sequence analysis of the genes (VENKATESAN et al. 1988a; see below) in agreement with the postulated cell surface exposure of the antigens (HROMOCKYJ and MAURELLI 1989; BUYSSE et al. 1990).

Three groups have independently determined the DNA sequence of various extents of the *ipa* gene region from *S. flexneri* 5 (BAUDRY et al. 1988; VENKATESAN et al. 1988a) and *S. flexneri* 2a (SASAKAWA et al. 1989), and a compilation of this data shows that the 8880-bp sequence encodes the synthesis of seven polypeptides (ADLER et al. 1989; BUYSSE et al. 1990; VENKATESAN and BUYSSE 1990; WATANABE et al. 1990). In the 8880-bp sequence only five base pair differences were found between *S. flexneri* 5 and *S. flexneri* 2a. The genes for these proteins are transcribed from the same sense strand of DNA, using all three reading frames, and are organized from the 5′ end as follows: 24-kDa ORF, 18-kDa ORF (*ippI*), *ipaB, ipaC, ipaD, ipaA,* and *virB* (*ipaR, invE*) (Fig. 1). The G + C content of the *ipa* genes averages 37%, considerably lower than the 50% G + C content of the *Shigella* chromosome. This finding is also reflected in other invasion plasmid genes that have been sequenced (*virF, virG, ipaH*) (SAKAI et al. 1986b; KATO et al. 1989; LETT et al. 1989; HARTMAN et al. 1990), and it has been postulated that the difference in G + C content of the plasmid and chromosome reflects different evolutionary origins for the respective replicons, or that the low G + C content of the plasmid genes makes them more accessible to transcriptional and associated regulatory proteins.

Plasmid subclones derived from the *ipa* gene region encode the synthesis of proteins in minicells or maxicells whose sizes correspond with the predicted molecular weights determined by DNA sequence analysis (BAUDRY et al. 1988; VENKATESAN et al. 1988a; SASAKAWA et al. 1989; BUYSSE et al. 1990). Thus, plasmid recombinant pHC17, containing a 4.7-kb *Hin*dIII insert, synthesized 18-kDa (*ippI*), 62-kDa (*ipaB*), and 42-kDa (*ipaC*) proteins (VENKATESAN et al. 1988a). A contiguous 8.0-kb *Eco*RI fragment (pEC14) overlaps the 3′ end of the pHC17 insert and directs the synthesis of the 37-kDa IpaD protein, the 70-kDa IpaA product, and the 34-kDa VirB(IpaR) protein; in addition, two proteins of 27 and 28 kDa are encoded downstream of the 3′ end of *virB*(*ipaR*), although their DNA sequence has not been determined (VENKATESAN et al. 1988a; BUYSSE et al. 1990). As can be seen from Fig. 1, the genes in region 2 are closely linked. Five nucleotides separate the *ippI* termination codon and the initiation codon of *ipaB*, and the *ipaD* and *ipaA* genes are separated by 8-bp. The *ipaC* and *ipaD* ORFs contain a 50-bp spacer, and the largest distance between consecutive *ipa* genes is the 459-bp that marks the end of *ipaA* and the beginning of *virB* (*ipaR*). The *ipaC* gene contains two in-frame ATG start sites, the first of which overlaps eleven amino acid residues at the 3′ end of the *ipaB* gene. Based on preliminary amino acid sequencing of the IpaC protein (SANKARAN et al. 1989), it seems likely that the second ATG start for *ipaC*, that is spaced 19 bp from the termination codon of *ipaB*, is the actual codon used for translation initiation.

The transcriptional organization of region 2 has been investigated by a variety of techniques, including Northern blot analysis, S1 nuclease protection experiments, and the use of promoter-deficient vectors to identify promoters derived from the *ipa* gene region (VENKATESAN et al. 1988a; SASAKAWA et al.

1989). Based on mutagenesis experiments using either Tn*5* or λ p*lac*Mu*53* in *S. flexneri* 2a or Tn*3-lac* in *S. sonnei* and Western blot analysis of the resulting mutants (BAUDRY et al. 1987; SASAKAWA et al. 1989; HROMOCKYJ and MAURELLI 1989; WATANABE et al. 1990), it is clear that insertions in the upstream 24-kDa ORF, *ippl*, or *ipaB* genes disrupt the expression of downstream loci such as *ipaC*, *ipaD*, and *ipaA*, while such insertions in *ipaC* or *ipaD* have little effect on the production of the 24-kDa protein, Ippl, and IpaB. This finding would support the idea that an operon, encompassing, in order, the 24-kDa, *ippl*, *ipaB*, *ipaC*, *ipaD*, and *ipaA* genes, is the regulatory model for *ipa* gene expression.

VENKATESAN et al. (1988a) used Northern blot analysis of *S. flexneri* 5 total RNA prepared from virulent cells grown at 30 °C and 37 °C to demonstrate that temperature regulation of *ipa* gene expression occurred at the level of transcription. Defined *ipa* gene segments carrying either *ipaB* or *ipaC* insert DNA were used to hybridize the RNA; the *ipaB* probe hybridized a 2.4-kb transcript, while the *ipaC* probe detected a 1.4-kb band. Both transcripts were detected when the 4.7-kb *Hin*dIII insert of pHC17, carrying *ippl–ipaB–ipaC* genes, was used as a radiolabeled probe. Overexposure of the Northern blots revealed a larger 3.3–3.5-kb transcript with all three probes. RNA hybridization with *ipaD* and *ipaD–ipaA* gene probes indicated the presence of at least two transcript species, 4.5 and 2.7 kb in size. While using probes corresponding to the 1.5-kb 24-kDa *ippl–ipaB′ EcoRI*, the 0.4-kb *ipaB′ Sal*I, and the 0.7-kb *ipaD′–ipaA′ Hin*dIII fragments, SASAKAWA et al. (1989) detected 7.5-kb, 4.5-kb, and 4.0-kb transcripts, all of which are eliminated in a 24-kDa ORF: : Tn*5* mutant. Three potential promoter sequences of region 2, called P_1, P_2, and P_3, were identified by subcloning the various restriction fragments into a promoter probe vector. Two of them were contained in 1.3-kb *Pst-Hin*dIII and 440-bp *Hin*dIII fragments overlapping the 5′ end of the 24-kDa ORF, and one was contained in a 451-bp *Nsi*I-*Hin*dIII fragment near the 5′ end of the *ipaD* gene (Fig. 1). S1 nuclease mapping fixed the initiation site for the 7.5-kb and the 4.0-kb transcripts 670 bp and 44 bp upstream of the 5′ end of the 24-kDa ORF, respectively. In summary, the evidence to date suggests that the *ipa* gene region consists of

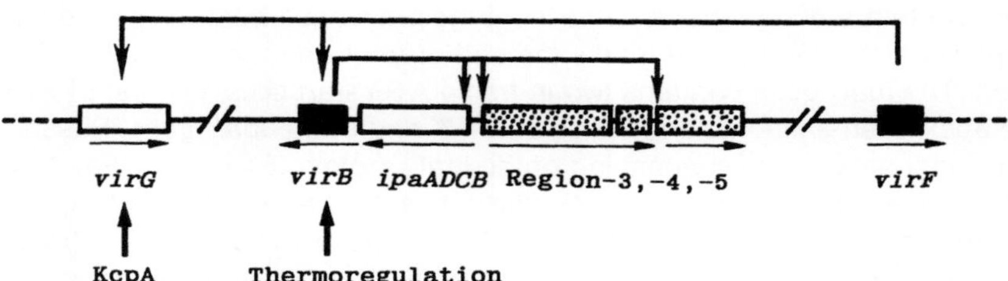

Fig. 2. Regulatory system of the plasmid-encoded virulence genes in *S. flexneri* (SAKAI et al. 1988; ADLER et al. 1989; PAL et al. 1989; TOBE et al., unpublished data; see Sects. 2.3, 2.4, 2.6, 3.1, and 3.2). The *arrows* beneath the virulence regions indicate the orientation of transcription

at least three operons, coordinately regulated by the *virB* gene product, through temperature-activated transcription (Fig. 2; see Sect. 2.4, 3.1, and 3.2). One operon consists of genes encoding the 24-kDa, Ippl, IpaB, and IpaC proteins. Open reading frames for IpaD and IpaA comprise a second transcriptional unit, with the *virB* (*ipaR*; see Sect. 2.4) gene being transcribed from a third and separate promoter. In addition, the 7.5-kb RNA transcript which covers the *ipa* gene region supplements full expression of the Ipa proteins (SASAKAWA et al. 1989).

Although IpaB, IpaC, and IpaD proteins have not been purified, the nucleotide sequence of the genes yields some information regarding their physical properties. Perhaps the most striking feature of these antigens is their marked hydrophilic nature. Of the four antigens, the IpaD protein has the most pronounced hydrophilicity with the IpaB, IpaC, and IpaA proteins consisting of hydrophilic domains at the amino and carboxy terminal ends and significant hydrophobic regions in the interiors of the molecules. The epitopes defined by MAbs raised against IpaB and IpaC correspond in their genetic map position with regions of hydrophilicity on both molecules (MILLS et al. 1988; VENKATESAN et al. 1988a). The proteins are also distinguished by their lack of cysteine residues (one in IpaA, IpaD, and IpaB; none in IpaC) and by the absence of signal peptide sequences at their amino termini. The latter observation is intriguing in view of the fact that the IpaB and IpaC proteins are cell surface-expressed polypeptides (MILLS et al. 1988; HROMOCKYJ and MAURELLI 1989). The degree of their integration, if any, with the outer membrane has not been determined, although it is likely that the antigens are only loosely affiliated with the outer envelope since the four antigens are extracted readily by washing the cells in distilled water. MAbs recognizing amino terminal epitopes of IpaB and IpaC bind to whole bacterial cells in an enzyme-linked immunosorbent assay, indicating that the antigens are expressed on the bacterial surface (MILLS et al. 1988; see Sect. 2.5). If the IpaB, IpaC, and IpaD antigens are critical ligands for the invasive phenotype, as is suggested by genetic data, then one might anticipate that binding of Ipa-specific MAbs would effect the cells invasive ability. In fact, the 2F1 IpaB-specific MAb partially inhibits the formation of plaques in BHK cell monolayers, while the IpaC-specific 2G2 MAb enhances plaque formation; both of these observations suggest that unencumbered 2F1 and 2G2 epitopes are needed for a wild-type level of invasion (MILLS et al. 1988).

A number of genetic clues indicate that a series of genes encoded in regions 3, 4, and 5 are critical for the proper post-transcriptional modification, transport, or presentation of the antigens on the cell surface. The associated phenotype has been designated Mxi$^+$ (*m*embrane e*x*pression of *I*pa antigens) or Spa$^+$ (*s*urface *p*resentation of Ipa *a*ntigens) (HROMOCKYJ and MAURELLI 1989; M. VENKATESAN et al., unpublished observations) and is based on the observation that *lacZ* gene fusions in loci 10–15 kb upstream of the *ipa* genes (HROMOCKYJ and MAURELLI 1989) or deletions that overlap virulence regions 3, 4, and 5 (BUYSSE et al., unpublished data) express normal levels of the Ipa antigens but that these antigens are not accessible to MAb binding or to other Ipa antigen-specific antisera. The determination of how the Ipa antigens interface with the cell surface and a more precise definition of "invasion epitopes" will add immensely to our understanding of the molecular aspects of colonic epithelial cell invasion by *Shigella*.

2.3 *virF*

The first plasmid-specified *vir* gene to be identified and sequenced was *virF* (SAKAI et al. 1986a, b). The *virF* gene was initially cloned from pMYSH6000 of *S. flexneri* 2a into *E. coli* K-12 as an essential determinant of the Pcr^+ phenotype and was later determined to be required for the Inv^+ phenotype as well. The defined *virF* coding region is a 1.0-kb DNA sequence in *Sal*I fragment F, located 40 kb from the distal end (3′) of region 5 on pMYSH6000 (Fig. 1). The cloned *virF* gene complements Inv^- Pcr^- mutants carrying small deletions in the *Sal*I F fragment, but does not complement more extensive deletions encompassing the 31-kb virulence DNA segment. PAYNE and FINKELSTEIN (1977) originally indicated that the Pcr^+ phenotype of *Shigella* correlated with the virulent phenotype. Subsequent studies have shown that the large plasmid of *Shigella* is required for the Pcr^+ as well as the Inv^+ phenotypes (MAURELLI et al. 1984a; DASKALEROS and PAYNE 1985; SASAKAWA et al. 1986a), and that both phenotypes are temperature regulated (MAURELLI et al. 1984a, b). Though the exact relationship between the ability to bind Congo red dye and invasion is not known, there is some indication that dye binding promotes efficient invasion of epithelial cells (DASKALEROS and PAYNE 1987). The expression of invasion plasmid antigens IpaA, IpaB, IpaC, IpaD, and VirG decreased dramatically in a *virF*::Tn5 mutant, and transcriptional analysis has indicated that the *virF* gene is a positive regulator of the *virG* and the *ipa* operons (SAKAI et al. 1988). Subsequent studies have shown that the expression of the *ipa* genes is regulated by *virF* through the transcriptional activation of a second positive regulator, *virB*, encoded at the 3′ end of the *ipa* operon (Fig. 2) (see the following section; ADLER et al. 1989).

The DNA sequence of the *virF* region encodes three proteins of 30 kDa, 27 kDa and 21 kDa as identified in minicell analysis (SAKAI et al. 1986a, b). The two smaller proteins are produced from in-frame translation start sites within the largest open reading frame for the 30-kDa protein. When a small DNA region containing only the first translation start site is deleted, VirF function is completely abolished together with the expression of the 30-kDa protein. This finding indicates that the regulatory function is encoded by the 30-kDa *virF* product consisting of 262 amino acid residues (SAKAI et al. 1988). The DNA sequence of the *virF* gene shows it to be rich in A and T (63%), a property shared with other plasmid-encoded *vir* genes (see the previous section; SAKAI et al. 1986b). Not surprisingly a *virF* probe from pMYSH6000 hybridizes the large plasmids of all virulent *Shigella* and EIEC strains (SAKAI et al. 1988) and explains the observation by KATO et al. (1989) that the DNA sequence of *virF* from pSS120 in *S. sonnei* is identical with that of *S. flexneri* 2a.

The 262-amino acid residue sequence of the 30-kDa VirF protein, deduced from the DNA sequence, has recently been shown to have a 36.2% sequence similarity with the primary sequence of CfaD (SAVELKOUL et al. 1990) and Rns (CARON et al. 1989), both of which are positive regulators for the expression of colonization factor antigens I (CFA/I) (EVANS et al. 1975) and II (CFA/II) (EVANS and EVANS 1978), respectively. Out of 53 amino acids between residues 203 and 255, 32 are shared with CfaD or Rns. This region of CfaD and Rns contains amino acid sequences common to other positive regulators as well, such as AraC, the regulatory protein of the arabinose operon in *E. coli* (SMITH and SCHLEIF 1978); ErC, the corresponding regulator from *Erwinia caratovora* (LEI et al. 1985); RhaR and RhaS, the regulator proteins of the rhamnose operon in *E. coli* (TOBIN and SCHLEIF 1987); and a distinct VirF, a regulatory protein for the *yop* genes in *Yersinia enterocolitica* (CORNELIS et al. 1989). Thus, the partial amino acid homology found in the C terminal region of the 30-kDa VirF protein may reflect a common regulatory mechanism that VirF shares with the AraC family of DNA-binding proteins.

2.4 *virB (invE or ipaR)*

The *virB* gene was initially recognized by random Tn*5* insertions as one of the virulence regions (region 1) within the 31-kb DNA segment of pMYSH6000 (see Sect. 2.1; SASAKAWA et al. 1988). This gene is now known to be a second positive regulator whose expression is governed by *virF* and temperature (ADLER et al. 1989; T. TOBE et al., unpublished results; see Sects. 2.3 and 3.1). The regulatory function of *virB* involves the activation of transcription of the *ipa* gene and other *inv* operons as determined by Northern blot hybridization. Mutations in the *virB* gene block transcription of the *ipa* genes, but do not affect *virF* or *virG* transcription, thus accounting for the avirulent Inv$^-$ Pcr$^-$ phenotype of the *virB* mutants (Sect. 2.1). The coding region for *virB* is located downstream of the *ipa* operon (Fig. 1), and the nucleotide sequence of the gene reveals an open reading frame encoding a 35.4-kDa protein comprising 309 amino acid residues whose molecular mass is close to the 33-kDa virB protein seen in minicells (ADLER et al. 1989).

A similar regulatory gene of pWR100 in *S. flexneri* 5 has recently been identified and designated *ipaR* (BUYSSE et al. 1990). The DNA sequence of the *ipaR* region establishes that the initiation codon for *ipaR* is 459 bp from the 3′ end of the *ipaA* gene (BUYSSE et al. 1990; VENKATESAN and BUYSSE 1990; see Sect. 2.2), the most downstream gene in the *ipa* regulon, and that *ipaR* also encodes a protein with 309 amino acid residues, as found for VirB. Comparison of the DNA sequences for *virB* and *ipaR* show identity through the length of the genes except for a transversion of T (*virB*) to C (*ipaR*) at bp 780 and bp 1278.

WATANABE et al. (1990) have confirmed the identity of this second positive regulator gene, designated *invE* in *S. sonnei*, whose expression is governed by the *virF* gene of pSS120. A 37-kb clone of pSS120, pJK1142, confers the Inv^+ phenotype upon a *S. sonnei* mutant lacking the large plasmid, but possessing a cloned *virF* gene (pJK1143) (Fig. 1). The analysis of Tn*3*-*lac*-generated *lacZ* fusions in pJK1142 showed that Tn*3*-*lac* insertion into the 1.0-kb *invE* region reduces Ipa protein expression resulting in an Inv^- phenotype. The DNA sequence of the *invE* region encodes a 35.4-kDa protein consisting of 309 amino acid residues identical to the VirB (IpaR) protein.

The calculated pI of this positive regulatory protein indicates a basic (pI 9.7), positively charged protein, and its hydrophobicity profile reveals a number of hydrophilic regions, as expected for a cytosolic protein. A striking homology exists between VirB (IpaR and InvE) and the related plasmid partition protein ParB of plasmid P1 (ABELES et al. 1985) and SopB of plasmid F (MORI et al. 1986). Though *invE* is not functionally interchangeable with *parB* (see WATANABE et al. 1990), the 42.8% identity over 278 amino acid residues may indicate that, like ParB or SopB, the VirB protein exerts its regulatory effect by direct binding of the protein to a specific DNA target or through protein–protein interaction with the transcriptional apparatus. Indeed, the purified InvE protein has recently been observed to bind to DNA segments of pJK1142 which code for promoter activity (ARAKAWA et al., unpublished data.)

2.5 Other Invasion-Associated Regions

As mentioned in Sect. 2.2, the functional organization of the *ipa* operon and the *virB* gene and the corresponding DNA sequence have been determined. Nevertheless, the remaining 20-kb DNA portion of the 31-kb DNA virulence segment corresponding to regions 3, 4, and 5 of pMYSH6000 (SASAKAWA et al. 1988) and the analogous 20-kb segments of pHS4108 (MAURELLI et al. 1985) and pJK1142 (KATO et al. 1989) have not been well characterized (see Fig. 1). The only locus described within this region is *invA*, which maps to contiguous 2.6- and 4.1-kb *Hin*dIII segments of the *S. sonnei* pSS120 virulence plasmid, although the exact genetic region of this locus remains to be determined (see Sect. 2.1; WATANABE and NAKAMURA 1986). Despite the paucity of genetic information for this region, the consensus is that this considerable extent of DNA codes for a variety of *vir* genes involved in the adherence of bacteria to the target cell surface and in the surface expression of the IpaB and IpaC proteins.

PAL and HALE (1989) have recently found that the adherence of *S. flexneri* strain M90T to HeLa cells at 4 °C is ten fold stronger than the isogenic plasmidless Inv^- mutant and that the surface of the Inv^+ parent is more hydrophobic than the mutant. The adherence reaction occurs at 4 °C in this wild-type strain grown at 37 °C, but not when it is grown at 30°C. Similarly, Inv^- mutants of *S. flexneri* with Tn*5* in the 20-kb portion of regions 3, 4, and 5 are all Pcr^- (see Sect. 2.1) and present a diminished capacity to adhere to the surface of LLC-MK2 cells (N. OKADA et al., unpublished data). The loss of adherence in these Tn*5* mutants is concomitant with a decrease in the bacterial surface net charge, as determined by binding to an anion-exchange matrix diethylaminoethanol (DEAE)-cellulose gel. These results strongly suggest that genes within the 20-kb DNA sequence affect the surface properties of *Shigella*.

It has recently been suggested that the 20-kb DNA portion has a role in the surface expression of IpaB and IpaC proteins (HROMOCKYJ and MAURELLI 1989; BUYSSE et al. 1990). *inv*::*lac* operon fusions in the 20-kb DNA region of pSfa140, the large plasmid of *S. flexneri* 2a strain 2457T, were examined for levels of β-galactosidase as well as for surface expression of IpaB and IpaC antigens in a whole-cell enzyme-linked immunosorbent assay (ELISA) using the 2F1 and 2G2 MAbs specific for IpaB and IpaC, respectively (MILLS et al. 1988; see Sect. 2.2). The results showed that the expression of the *inv* operon, determined by the *lac* fusion, is temperature regulated and that the surface-expressed IpaB and IpaC antigens are decreased to less than half of those found in the wild-type parent. It has also been shown by BUYSSE et al. (1990) that reconstitution of M90T-A_3, a spontaneous regions 3, 4, and 5 deletion mutant of *S. flexneri*, with cloned segments of the *virB* (*ipaR*) and *ipaBCD* gene is insufficient to restore the invasion phenotype since the produced Ipa antigens are not detected on the bacterial cell surface. These observations raise the possibilities that genes encoded by the 20-kb virulence DNA portion are involved in (a) post-translational modification of the Ipa proteins, or (b) transfer and positioning of the Ipa proteins in the cell envelope of the bacterium (see Sect. 2.2).

2.6 *virG (icsA)*

The *virG* gene is located on *Sal*I fragment G of pMYSH6000 (Fig. 1) (SASAKAWA et al. 1986b). Avirulent mutants with Tn*5* insertions in the 3.6-kb *virG* locus retain the invasion and Congo red binding phenotypes, thus distinguishing these mutants from other avirulent plasmid mutations that abolish invasion (see Sect. 2.1). Though capable of invading epithelial cells, *virG* mutants accumulate in the cytoplasm of the target cell and do not spread to adjacent cells, suggesting that the *virG* locus is required for inter- and intracytoplasmic movement of the bacteria but not for intracellular multiplication (MAKINO et al. 1986; LETT et al. 1989). Minicell analysis of *virG* plasmid recombinants revealed that the 3.6-kb *virG* region expresses eight polypeptides ranging in size from 130 kDa to 25 kDa (LETT et al. 1989). The largest 130-kDa protein is recognized by convalescent-phase antiserum of shigellosis patients (SAKAI et al. 1988). Genetic complementation of the *virG* region suggests that the determinant consists of a single cistron, and the nucleotide sequence indicates that the largest open reading frame codes for a protein of 1102 amino acid residues with a molecular mass of 116.4 kDa. Extrinsic radioiodination of whole bacterial cells shows that this polypeptide is exposed on the surface of *Shigella* (LETT et al. 1989). As was found for other virulence-associated genes, hybridization studies show that sequences homologous with *virG* are conserved among other large plasmids of *Shigella* and EIEC (MAKINO et al. 1986).

A 130- to 140-kDa protein, which is the *virG* gene product (SAKAI et al. 1988; LETT et al. 1989), was initially identified as one of the plasmid-encoded antigens specifically recognized by convalescent-phase human or monkey sera (OAKS et al. 1986). BERNARDINI et al. (1989) have recently identified a pWR100 virulence locus in a 6.2-kb *Sal*I-*Eco*RI fragment of *S. flexneri* 5 strain M90T similar to the *virG* region of pMYSH6000 and have shown that the locus expresses a 120-kDa antigen. This antigen is responsible for a unique property of invading *Shigella*, namely the localized deposition of F-actin trailing one pole of the bacterial cell and extending in a filament through the host epithelial cytoplasm. The observed trail of F-actin in the infected epithelial cells apparently provides a motive force since the intracellular spread of invading *Shigella* is blocked by the addition of cytochalasin D (HALE et al. 1979), an inhibitor of actin polymerization. It is postulated that the accumulation of F-actin fibrils results in the formation of extracellular protrusions through which bacteria could penetrate adjacent cells. Indeed, on the surface of infected cells, extracellular protrusions which include the bacteria and associated trails of F-actin are observed (see also PAL et al. 1989). Accordingly, BERNARDINI et al. (1989) have named the *virG* analogue *icsA* (*i*ntercellular *s*pread). It should be noted that a similar mechanism has been documented in the intercellular spread of *Listeria monocytogenes* through macrophage monolayers (TILNEY and PORTNOY 1989).

The expression of the *virG* gene is regulated at the transcriptional level by *virF* (SAKAI et al. 1988; see Sect. 2.3). In addition, the expression of a large antigen of 130–140 kDa (presumably VirG) encoded by pWR100 of *S. flexneri* 5 has recently been shown to be under the regulatory control of the chromosomal *kcpA* locus (Fig. 2) (PAL et al. 1989). In fact, *virG* and *kcpA* mutants display similar phenotypes in that both are Inv^+ and Pcr^+, but are incapable of intracellular and intercellular spread, resulting in a negative Sereny test phenotype (SANSONETTI et al. 1982; MAKINO et al. 1986; BERNARDINI et al. 1989; PAL et al. 1989; YAMADA et al. 1989). It is not known whether *kcpA* regulates the expression of *virG* directly or in conjunction with the *virF* function.

2.7 *ipaH*

As described in Sect. 2.2, the λgt11 cloning system was used by BUYSSE et al. (1987) to create an expression library of invasion plasmid antigen genes from *S. flexneri* 5. In the course of characterizing 28 putative λgt11*ipaB* recombinants, the authors noted that a subset of 17 clones carried insert DNA that did not hybridize with the insert DNA of the remaining 11 recombinants that synthesized either complete or versions truncated of β-galactosidase fusion versions of the 62-kDa IpaB antigen. Thus, it appeared that this new class of λgt11Sf1 recombinants directed the synthesis of an antigen similar in size to the IpaB protein, but distinct as determined by DNA sequence homology measurements. Further DNA hybridization studies revealed that this newly isolated antigen gene was present in multiple copies (five) on the pWR100 invasion plasmid, in sharp contrast to the unit copy representation found for the *ipaB, ipaC, IpaD,* and *ipaA* genes (HARTMAN et al. 1990). This multicopy antigen gene has been named *ipaH*, and the five copies of the *ipaH* gene isolated from pWR100 are designated $ipaH_{9.8}$, $ipaH_{7.8}$, $ipaH_{4.5}$, $ipaH_{2.5}$, and $ipaH_{1.4}$, the subscripts for each *ipaH* allele referring to the molecular sizes (in kilobases) of the *Hin*dIII fragments that contain the genes.

Affinity-purified antibodies prepared from representative λgt11*ipaB* and λgt11*ipaH* recombinants reacted with an approximately 60-kDa protein present in whole cell lysates of virulent *S. flexneri* 5, but IpaH affinity-purified antibodies did not recognize IpaB antigen and vice versa (BUYSSE et al. 1987). This result demonstrated that the IpaB and IpaH antigens are immunologically distinct. The IpaH antigen is expressed during *Shigella* infections since antibody recognizing this protein has been detected in convalescent human and monkey antisera (HARTMAN et al. 1990). Unlike other plasmid virulence genes, expression of *ipaH* is independent of temperature regulation or induction by *virF* or *virB*; to date, transcripts and protein products have been detected in *S. flexneri* 5 from both the $ipaH_{7.8}$ and $ipaH_{4.5}$ genes (M. VENKATESAN et al., unpublished observations). Although the conservation of *ipaH* in multiple copies suggests a powerful selection for this gene (s), the presence of these copies has complicated the construction of $IpaH^-$ mutants and so precluded the identification of a virulence-associated IpaH phenotype.

Nucleotide sequence analysis of an $IpaH^+$ 2.9-kb fragment (pWR390) of the pWR100 plasmid has shown that the $ipaH_{7.8}$ gene encodes a 60.8-kDa, 532-amino acid residue protein that is detected in maxicells and is an abundant immunogen produced by the M90T strain of *S. flexneri* 5 (HARTMAN et al. 1990). The hydropathy index of the IpaH protein indicates a predominantly hydrophilic polypeptide with a few small hydrophobic domains, mostly concentrated in the amino terminal end of the molecule. The polypeptide does not contain an amino terminal signal peptide sequence, a property it shares with other *ipa* gene products as well (see Sect. 2.2). Analysis of the amino acid sequence of the IpaH protein revealed six evenly spaced 14-residue repeat motifs consisting of Leu-X_2-Leu-Pro-X-Leu-Pro-X_2-Leu-X_2-Leu (where X represents any amino acid) located between residues 39 and 149 in the amino terminal end of the molecules. Each repeat of this motif is separated from the next by six amino acids, the fifth of which is a conserved asparagine residue. The regular spacing of leucine residues in this motif suggests that a uniform hydrophobic surface may be presented on one side of an α-helical array of the residues. This arrangement might facilitate oligomerization of IpaH molecules or interactions with other proteins that present a similar hydrophobic configuration. An interesting observation has been that the leucine-rich LPX motif of IpaH is also found in the YopM protein of *Yersinia pestis*, and both proteins share homology with the GP1b glycoprotein of human platelets (M. VENKATESAN and J. BUYSSE, manuscript submitted; LEUNG et al. 1990), implying that interference with platelet-mediated events may be the mechanism by which IpaH contributes to *Shigella* virulence.

A total of 427 bp separate the termination codon of *ipaH*$_{7.8}$ from a second open reading frame on pWR390 (HARTMAN et al. 1990). Complete sequencing of this second ORF revealed the same set of LPX motifs, reiterated nine times instead of six, as was found in *ipaH*$_{7.8}$, and complete DNA sequence homology with the middle and carboxy terminal portions of *ipaH*$_{7.8}$. The second ORF was thus designated *ipaH*$_{4.5}$, and together these tandemly repeated *ipaH* genes have been mapped to *Sal*I fragment B between *virG* and *virB* loci (M. VENKATESAN and J. BUYSSE, manuscript submitted). Two other copies of pWR100 *ipaH* genes (*ipaH*$_{2.5}$ and *ipaH*$_{1.4}$) have been cloned and sequenced and, although they do not contain the LPX motifs, they are completely homologous with *ipaH*$_{7.8}$ over the 590-bp segment in the middle of the gene.

The Southern DNA hybridization pattern of *ipaH* genes observed in pWR100 is unique in that the same pattern is not detected in other *Shigella* invasion plasmids. Virulence plasmids from representative serotypes of *S. flexneri*, *S. boydii*, *S. dysenteriae*, *S. sonnei*, and EIEC all contain *ipaH* genes arranged in distinctive patterns characteristic of the given serotype. This heterogeneity and the demonstrated *Shigella* specificity of *ipaH* gene probes (VENKATESAN et al. 1989) combine to make this gene a powerful genetic probe for both diagnostic and epidemiologic work. Recently, copies of *ipaH* have been detected on the chromosome of invasive and noninvasive shigellae; this finding should allow the use of genomic *ipaH* RFLP analysis to follow, geographically, the dissemination of a particular *Shigella* strain through a susceptible population (J. BUYSSE et al., manuscript in preparation).

3 Regulation of Expression

3.1 Positive Control of *inv* Genes

The expression of the invasion phenotype in *Shigella* is under the control of a dual activation system by *virF* and *virB* (ADLER et al. 1989; see Sects. 2.2–2.4). In addition, epithelial cell penetration is subject to thermoregulation through the negative control of the chromosomal *virR* locus (MAURELLI and SANSONETTI 1988; see Sect. 3.2).

VirB mutants of *S. flexneri* 2a do not produce detectable levels of Ipa proteins in Western blot analysis, with the exception of the IpaH protein. When the cloned *virB* region is introduced into *virB* mutants, the invasion phenotype as well as the expression of Ipa proteins is restored. The transcription of the 1.1-kb *virB-m*RNA does not occur in *virF* mutants, but does resume when the *virF* gene is reintroduced (ADLER et al. 1989). Whereas the transcription of *ipa* and *inv* operons in the *virB* mutant decreases to nondetectable levels, those of the *virF* and *virG* genes do not (T. TOBE, unpublished data). Based on these results, ADLER et al. (1989) have proposed that the expression of the *ipa* and *inv* operons on the large plasmid is under the control of the dual activation system directed by *virF* and *virB*.

A similar activation system was subsequently found on pSS120 in *S. sonnei* (WATANABE et al. 1990). The level of *β*-galactosidase expressed from the *invE*::Tn*3*-lac fusion increases in accord with the dose of the *virF* gene. Expression of the *lacZ* gene fused with *inv* operons occurs only when both *virF* and *invE* genes exist, further supporting the dual activation system proposed by ADLER et al. (1989).

By overexpressing the *virB* gene in a *virF* mutant of *S. flexneri* 2a, the direct role of VirB activation of *ipa* and *inv* operon expression can be demonstrated (T. TOBE et al., unpublished data). Linking the *tac* promoter to *virB* causes the activation of the *ipa* and *inv* operons and consequently releases the operons from temperature-dependent expression since invasion can then be achieved at 30 °C. However, as expected, the activated *tac-virB* construct present in a *virF* mutant does not confer a positive plaque assay or Sereny test even at 37 °C (T. TOBE et al., unpublished data). The inability to restore a wild-type virulence level in the *virF* mutant with *tac-virB* can be accounted for by the lack of VirG function, for which VirF is required (SAKAI et al. 1988), thus confirming the regulatory circuit proposed by ADLER et al. (1989) (Fig. 2).

3.2 Thermoregulation of *inv* Genes

The best-known environmental signal for the regulation of the virulence phenotype of *Shigella* is temperature. *Shigella* strains which are invasive when grown at 37 °C become noninvasive when grown at 30 °C. This phenotypic change is reversible, and the restoration of invasion capacity requires de novo protein synthesis (MAURELLI et al. 1984b). Thus, the expression of *inv* genes themselves and/or that of the regulatory genes, such as *virF* or *virB*, appears to be subject to thermoregulation.

MAURELLI and SANSONETTI (1988) hypothesized that the expression of *inv* genes on the large plasmid was repressed at 30 °C by a *trans*-acting regulator such as a repressor protein. To characterize this regulator, the authors first isolated *inv::lacZ* operon fusion on the large plasmid of *S. flexneri* 2a (2457T) whose Lac$^+$ phenotype was expressed at 37 °C but not at 30 °C, a construction in which the expression of the *lacZ* gene was assumed to be under the control of a temperature-regulated *inv* operon. They then undertook Tn*10* mutagenesis of the *inv::lacZ* fusions in order to isolate mutants able to express the Lac$^+$ phenotype at both 37 °C and 30 °C. One Tn*10* insertion mutation resulting in deregulation of the Lac$^+$ phenotype was then transduced to the wild-type strain and the transductant tested for its ability to invade HeLa cells at 30 °C. The site of the Tn*10* insertion carried by the transductant was near the *galU* gene at 28′ in the *Shigella* chromosome, and the determinant was designated *virR*.

The expression of an Inv$^+$ phenotype by a *virR* mutant at 30 °C coincides with the expression of the Ipa proteins at 30 °C, indicating that the *virR* locus controls the temperature-regulated Inv$^+$ phenotype of *Shigella* through its effect on the *ipa* operon. The analysis of the effect of the *virR* mutation on the expression of other *inv* operons of the 2457T invasion plasmid reveals that the *virR* locus is involved in the expression of the *inv* genes at the transcriptional level (HROMOCKJY and MAURELLI 1989).

DORMAN et al. (1990) have recently examined the role of *virR* function in the thermoregulation of *inv* gene expression on pSf140 in strain 2457T and indicate that the *virR* gene (MAURELLI and SANSONETTI 1989) is equivalent to the *osmZ* gene of *E. coli* (HIGGINS et al. 1988), which has previously been shown to mediate its regulatory effect through changes in DNA supercoiling. The proposed model implies that a change in temperature causes altered conformational properties of DNA sequences encoding the expression of the *inv* genes.

Recent studies have examined which *inv* gene on the large plasmid mediates the thermoregulation of a subset of *inv* genes on pMYSH6000 (T. TOBE et al., unpublished data). The transcriptional analysis of the effect of temperature on the activation of the two regulator genes, *virF* and *virB*, shows that the activation of *virB* strongly depends upon temperature compared with that of *virF*. The activation of *virB*, by increasing *virF* transcription, is achieved at 37 °C much more efficiently than at 30 °C. In contrast, levels of transcription of *ipa* and the other two *inv* operons expressed under the different levels of *virB* transcription are not affected by temperature at all. The activation of the *virB* gene through the construction of a constitutive *tac-virB* fusion leads to deregulation of the temperature-dependent invasion phenotype of *Shigella* (see Sect. 3.1), indicating that the temperature-regulated Vir$^+$ phenotype is mediated directly through the transcriptional activation of the *virB* gene on the large plasmid.

The above results are compatible with other observations. For instance, the expression of the Ipa proteins is temperature regulated in the 44-kb cosmid clone of pWR100, pHS4108 (MAURELLI et al. 1985), which codes for *virB* but not *virF* as judged by its physical map (see Fig. 1; SASAKAWA et al. 1988), and *lacZ* expression from the *inv*::Tn*3-lac* fusion in pSS120 shows temperature and VirF dependency (WATANABE et al. 1990). Although it is not understood how the *virB* gene is activated by the two factors (i.e., temperature and *virF*), the identity of the mediator gene for the temperature-regulated expression of *inv* genes on the large plasmid will facilitate the molecular study of how environmental signals such as temperature are transmitted to a set of genes.

3.3 Other Chromosomal Loci Affecting the Invasion Phenotype

Because of the close genetic relatedness between *E. coli* K-12 and *Shigella*, intergeneric conjugation between *E. coli* K-12 and *S. flexneri* has allowed classic genetic analysis of the virulence-associated chromosomal loci of

S. flexneri (FALKOW et al. 1963; FORMAL et al. 1971; GEMSKI et al. 1972; SANSONETTI et al. 1983b). On the chromosome, the genetic locus for the synthesis of the group-specific O antigen 3, 4 of *S. flexneri* is linked to the his region. A locus designated *kcpA*, for *k*eratoconjunctivitis *p*rovocation, maps near the *purE* gene. The *xyl-mtl* region is also known to be essential for provoking a positive Sereny reaction (FORMAL et al. 1965).

Recently the *iuc* gene, which encodes the hydroxamate aerobactin siderophore, has been implicated in producing a delayed positive Sereny test and was shown to affect extracellular growth within infected tissues (LAWLOR et al. 1987; NASSIF et al. 1988; SANSONETTI and ARONDEL 1989). The *sodB* gene, which encodes the iron-containing superoxide dismutase, has been shown to contribute to the survival of invading *Shigella* in mouse peritoneal macrophages or human polymorphonuclear leukocytes (FRANZON et al. 1990).

Although these bacterial genes do not directly influence invasion and spread, there are some chromosomal loci that do, most notably the *virR* gene (MAURELLI and SANSONETTI 1988) and the *kcpA* locus, which effects the expression of the *virG*(*icsA*) gene (PAL et al. 1989).

There are other, as yet undefined, chromosomal determinants that may impinge on the cells ability to infect and survive in host epithelial cells. A colonial variant of *S. flexneri* strain 2457T loses both the invasion phenotype and glycerol kinase activity encoded by the *glpk* gene (KIM and CORWIN 1974). The colonial variant strain, 2457O, fails to express some temperature-regulated proteins including IpaA, IpaB, IpaC, and IpaD and the 53-kDa, 25-kDa and 20-kDa proteins (HALE et al. 1985). An earlier genetic study (KIM and CORWIN 1974) revealed that transduction of the *glpK* gene from 2457T into the avirulent derivatives can restore the Inv^+ phenotype in roughly 50% of the transductants, whereas the same gene from *E. coli* cannot. It remains unknown how the *glpK* region of *S. flexneri* 2a is different from that of *E. coli* K-12; however, the DNA region linked to the *glpK* gene seems to code for a regulatory gene required for the expression of Ipa proteins and other invasion proteins.

Very recently, random Tn*5* insertions in *S. flexneri* 2a strain YSH6000 have been screened for mutants which were avirulent or less virulent by the focus plaque assay (OKADA et al. 1991). Among over 9000 independent insertion mutants, 50 mutants with single Tn5 insertions in their chromosome were assigned to 19 *Not* I fragments of the chromosomal DNA, designated A–S, according to their sizes. The 50 mutants were characterized with respect to their virulence phenotypes and were found to comprise three different mutations that affect invasion of epithelial cells, bacterial metabolism, and structure of lipopolysaccharides. Two mutants with reduced invasion ability were defective in the levels of IpaB, C, and D antigens produced as well as the corresponding mRNA. Assignment of the two Tn*5* insertions in the mutants to the *Not* I fragment map indicates that the two determinants are located in separate *Not* I fragments, C and M. These findings, together with the above examples, indicate that complete pathogenicity of *S. flexneri* requires various genes dispersed around the chromosome as well as on the large plasmid, and that the expression of some of them is under the control of a complicated regulatory network.

4 Conclusion

The invasion of epithelial cells by *Shigella* is a complicated process, involving numerous genes on the large virulence plasmid and on the chromosome. The full characterization of the corresponding *vir* genes is still in progress; however, it has become clear that *vir* genes on the large plasmids are involved not only in the invasion step, but also in the other various important stages in *Shigella* infection, as listed in Sect. 2.1. It is apparent from molecular studies that *vir* genes on the invasion plasmid form one or more regulons whose expression is under the control of a complex regulatory system, again involving many genes encoded by the large plasmid and the chromosome. For example, as seen in the temperature-regulated expression of the IpaB, IpaC, and IpaD proteins, these gene products are produced in defined amounts at critical times in the infection cycle in response to environmental signals such as temperature, osmolarity, or other biophysical factors. Thus, it is through the inheritance of the invasion plasmid and tightly regulated expression of the cognate *vir* genes that *Shigella* take advantage during infection of the human colonocyte intracellular environment.

As reviewed in this and in other chapters, it is clear that *Shigella* pathogenesis requires a large assembly of virulence factors that are not simultaneously needed during all stages of infection. Therefore, it will become important to accumulate knowledge about the molecular basis of function for each virulence protein as well as about the properties of virulence genes encoded by the large plasmid and the chromosome. Although investigation on the molecular pathogenesis of *Shigella* is relatively advanced, a number of questions remain regarding the molecular basis of certain steps in the infection process, including the following:

1. What are the biologic roles of the three Ipa proteins in the invasion of epithelial cells by *Shigella*, and what are the biochemical properties of the proteins? These questions must be addressed using biologically active and biochemically pure Ipa proteins.
2. What kinds of genes are encoded by virulence regions 3, 4, and 5, and which gene products are involved in the surface expression of IpaB, C, and D proteins and in the adherence properties of *Shigella* (see Sects. 2.3 and 2.7)?
3. What are the molecular bases of the regulatory function of the two positive regulators *virF* and *virB*? Investigation along this line should include biochemical studies on the two proteins, together with the identification of the target DNA sequence(s) recognized by these putative DNA-binding proteins.
4. How does the chromosomal *virR* gene regulate in a negative manner the expression of the *vir* genes on the large plasmid? Does it regulate the plasmid-encoded gene(s) directly or indirectly, with or without the involvement of other regulatory element(s)?

5. How is the expression of the *virG* gene regulated by the *kcpA* region on the chromosome?
6. What genes other than *virR* and *kcpA* on the chromosome are involved in the expression of *vir* genes on the large plasmid? Several genetic regions on the chromosome have been shown to affect the invasion capacity of *Shigella*, as described in the previous section.
7. What particular segments of the *virG* (*iscA*) protein are directly involved in the spread of invading bacteria through the aggregation of F-actin in the cytoplasm of epithelial cells?
8. What is role of IpaH proteins in *Shigella* infection, and at what stage of infection are they important? Why do the determinants exist as multiple copies both on the large plasmid and on the chromosome?
9. What kinds of environmental signals play a role in inducing the expression of *vir* genes on the large plasmid. How are they required for various stages of infection, and how is the signal transmitted to the target virulence gene?

Acknowledgements. The authors are grateful to Drs. M.M. Venkatesan and B. Adler for critical reading of the manuscript. C.S. is grateful to M. Yoshikawa for his encouragement in preparing this chapter.

References

Abeles AL, Friedman SA; Austin SJ (1985) Partition of unit-copy miniplasmids to daughter cells. III. The DNA sequence and functional organization of the P1 partition region. J Mol Biol 185: 261–272

Adler B, Sasakawa C, Tobe T, Makino S-I, Komatsu K, Yoshikawa M (1989) A dual transcriptional activation system for the 230 kilobase plasmid genes coding for virulence-associated antigens of *Shigella flexneri*. Mol Microbiol 3: 627–635

Baudry B, Maurelli AT, Clerc P, Sadoff JG, Sansonetti PJ (1987) Localization of plasmid loci necessary for the entry of *Shigella flexneri* into HeLa cells, and characterization of one locus encoding four immunogenic polypeptides. J Gen Microbiol 133: 3403–3413

Baudry B, Kaczorek M, Sansonetti PJ (1988) Nucleotide sequence of the invasion plasmid antigen B and C genes (*ipaB* and *ipaC*) of *Shigella flexneri*. Microb Pathog 4: 345–357

Bernardini ML, Mounier J, d'Hauteville H, Coquis-Rondon M, Sansonetti PJ (1989) Identification of *icsA*, a plasmid locus of *Shigella flexneri* that governs bacteria intra- and intercellular spread through interaction with F-actin. Proc Natl Acad Sci USA 86: 3867–3871

Buysse JM, Stover CK, Oaks EV, Venkatesan M, Kopecko DJ (1987) Molecular cloning of invasion plasmid antigen (*ipa*) genes from *Shigella flexneri*: analysis of *ipa* gene products and genetic mapping. J Bacteriol 169: 2561–2569

Buysse JM, Venkatesan MM, Mills JA, Oaks EV (1990) Molecular characterization of a *trans*-acting, positive effector (*ipaR*) of invasion plasmid antigen synthesis in *Shigella flexneri* sero type 5. Microb Pathog 8: 197–211

Caron J, Coffield LM, Scott JR (1989) A plasmid-encoded regulatory gene, *rns*, required for expression of CS1 and CS2 adhesins of enterotoxigenic *Escherichia coli*. Proc Natl Acad Sci USA 86: 963–967

Clerc PL, Ryter A, Mounier J, Sansonetti PJ (1987) Plasmid-mediated early killing of eucaryotic cells by *Shigella flexneri* as studies by infection of J774 macrophages. Infect Immun 55: 521–527

Cornelis GR, Biot T, Lambert de Rouvroit C, Michiels T, Mulder B, Sluiters C, Sory M-P, van Bouchaute M, Vanooteghem J-C (1989) The *Yersinia yop* regulon. Mol Microbiol 3: 1455–1459

Daskaleros PA, Payne SM (1985) Cloning the gene for Congo red binding in *Shigella flexneri.* Infect Immun 48: 165–168

Daskaleros PA, Payne SM (1987) Congo red binding phenotype is associated with hemin binding and increase infectivity of *Shigella flexneri* in the HeLa cell model. Infect Immun 55: 1343–1398

Dorman JC, Bhriain NN, Higgins CF (1990) DNA supercoiling and environmental regulation of virulence gene expression in *Shigella flexneri.* Nature 344: 789–792

Evans DG, Evans DJ (1978) New surface-associated heat-labile colonization factor antigen (CFA/II) produced by enterotoxigenic *Escherichia coli* of serogroups 06 and 08. Infect Immun 21: 683–647

Evans DG, Silver RP, Evans DJ, Chase DG, Gorbach SL (1975) Plasmid-controlled colonization factor associated with virulence in *Escherichia coli* enterotoxigenic for humans. Infect Immun 12: 656–667

Falkow S, Schrider H, Baron LS, Formal SB (1963) Virulence of *Escherichia–Shigella* genetic hybrids in the guinea pig. J Bacteriol 86: 1251–1258

Formal SB, LaBrec EH, Palmer A, Falkow S (1965) Restoration of virulence to a strain of *Shigella flexneri* by mating with *Escherichia coli.* J Bacteriol 80: 835–838

Formal SB, Gemski Jr P, Baron LS, LaBrec EH (1971) A chromosomal locus which controls the ability of *Shigella flexneri* to evoke keratoconjunctivitis. Infect Immun 3: 73–79

Franzon VL, Arondel J, Sansonetti PJ (1990) Contribution of superoxide dismutase and catalase activities to *Shigella flexneri* pathogenesis. Infect Immun 58: 529–535

Gemski Jr P, Sheahan DG, Washington O, Formal SB (1972) Virulence of *Shigella flexneri* hybrids expressing *Escherichia coli* somatic antigens. Infect Immun 6: 104–111

Hale TL, Morris RE, Bonventre PF (1979) *Shigella* infection of Henle intestinal epithelial cells: role of the host cell. Infect Immun 24: 887–894

Hale TL, Sansonetti PJ, Schad PA, Austin S, Formal SB (1983) Characterization of virulence plasmids and plasmid-associated outer membrane proteins in *Shigella flexneri*, *Shigella sonnei* and *Escherichia coli.* Infect Immun 40: 340–350

Hale TL, Oaks EV, Formal SB (1985) Identification and antigenic characterization of virulence-associated, plasmid-coded proteins of *Shigella* spp. and enteroinvasive *Escherichia coli.* Infect Immun 50: 620–629

Harris JR, Wachsmuth IK, Davis BR, Cohen ML (1982) High molecular weight plasmid correlates with *Escherichia coli* enteroinvasiveness. Infect Immun 37: 1295–1298

Hartman AB, Venkatesan M, Oaks EV, Buysse JM (1990) Sequence and molecular characterization of a multicopy invasion plasmid antigen gene, *ipaH*, of *Shigella flexneri.* J Bacteriol 172: 1905–1915

Higgins CF, Dorman CJ, Stirling DA, Waddell L, Booth IR, May G, Bremer E (1988) A physiological role for DNA supercoiling in the osmotic regulation of gene expression in *S. typhimurium* and *E. coli.* Cell 52: 569–584

Hromockyj AE, Maurelli AT (1989) Identification of *Shigella* invasion genes by inoculation of temperature-regulated *inv::lacZ* operon fusions. Infect Immun 57: 2963–2970

Ish-Horowicz D, Burke JF (1981) Rapid and efficient cosmid vector cloning. Nucleic Acids Res 9: 2989–2998

Kato J-I, Ito K, Nakamura A, Watanabe H (1989) Cloning of regions required for contact hemolysis and entry into LLC-MK2 cells from *Shigella sonnei* form I plasmid: *virF* is a positive regulator for these phenotypes. Infect Immun 57: 1391–1398

Kim R, Corwin LM (1974) Mutation in *Shigella flexneri* resulting in loss of ability to penetrate HeLa cells and loss of glycerol kinase activity. Infect Immun 9: 916–923

Kopecko DJ, Washington O, Formal SB (1980) Genetic and physical evidence for plasmid control of *Shigella sonnei* form I cell surface antigen. Infect Immun 29: 207–214

Kopecko DJ, Baron LS, Buysse JM (1985) Genetic determinants of virulence *Shigella* and dysenteric strains of *Escherichia coli*: their involvement in the pathogenesis of dysentery. Curr Top Microbiol Immunol 118: 71–95

LaBrec EH, Schneider H, Magnani JJ, Formal SB (1964) Epithelial cell penetration as an essential step in the pathogenesis of bacillary dysentery. J Bacteriol 88: 1503–1518

Lawlor KM, Daskaleros PA, Robinson RE, Payne SM (1987) Virulence of iron transport mutants of *Shigella flexneri* and utilization of host iron compounds. Infect Immun 55: 594–599

Lei S-P, Lin H-C, Heffernan L, Wilcox G (1985) *araB* gene and nucleotide sequence of the *araC* gene of *Erwinia carotovora.* J Bacteriol 164: 717–722

Lett M-C, Sasakawa C, Okada N, Sakai T, Makino S, Yamada M, Komatsu K, Yoshikawa M (1989) *virG*, a plasmid-coded virulence gene of *Shigella flexneri*: identification of the *virG* protein and determination of the complete coding sequence. J Bacteriol 171: 353–359

Leung KY, Reisner BS, Straley SC (1990) YopM inhibits platelet aggregation and is necessary for virulence of *Yersinia pestis* in mice. Infect Immun 58: 3262–3271

Makino S-I, Sasakawa C, Komatsu K, Kurata T, Yoshikawa M (1986) A genetic determinants required for continuous reinfection of adjacent cells on a large plasmid in *Shigella flexneri* 2a. Cell 46: 551–555

Maurelli AT, Sansonetti PJ (1988) Identification of a chromosomal gene controlling temperature-regulated expression of *Shigella* virulence. Proc Natl Acad Sci USA 85: 2820–2824

Maurelli AT, Blackmon B, Curtiss R III (1984a) Loss of pigmentation in *Shigella flexneri* is correlated with loss of virulence and virulence-associated plasmid. Infect Immun 43: 397–401

Maurelli AT, Blackmon B, Curtiss R III (1984b) Temperature-dependent expression of virulence genes in *Shigella* species. Infect Immun 43: 195–201

Maurelli AT, Baudry B, d'Hauteville H, Hale TL, Sansonetti PJ (1985) Cloning of plasmid DNA sequences involved in invasion of HeLa cells by *Shigella flexneri*. Infect Immun 49: 164–171

Mills JA, Buysse JM, Oaks EV (1988) *Shigella flexneri* invasion plasmid antigens B and C: epitope location and characterization by monoclonal antibodies. Infect Immun 56: 2933–2941

Mori H, Kondo A, Oshima A, Ogura T, Hiraga S (1986) Structure and function of the F plasmid genes essential for partitioning. J Mol Biol 192: 1–15

Murayama SY, Sakai T, Makino S-I, Kurata T, Sasakawa C, Yoshikawa M (1986) The use of mice in the Sereny test as a virulence assay of shigellae and enteroinvasive *Escherichia coli*. Infect Immun 51: 696–698

Nassif W, Mazert MC, Mounier J, Sansonetti PJ (1988) Evaluation with an *iuc*::Tn*10* mutant of the role of aerobactin production in the virulence of *Shigella flexneri*. Infect Immun 55: 1963–1969

Oaks EV, Hale TL, Formal SB (1986) Serum immune response to *Shigella* protein antigens in rhesus monkey and humans infected with *Shigella* spp. Infect Immun 48: 124–129

Okada N, Sasakawa C, Tobe T, Yamada M, Nagai S, Talukder KA, Komatsu K, Kanegasaki S, Yoshikawa M (1991) Virulence-associated chromosomal loci of *Shigella flexneri* identified by random Tn*5* insertion mutagenesis. Mol Microbiol 5: 187–195

Pal T, Hale TL (1989) Plasmid-associated adherence of *Shigella flexneri* in a HeLa cell model. Infect Immun 57: 2580–2582

Pal T, Newland JW, Tall BD, Formal SB, Hale TL (1989) Intracellular spread of *Shigella flexneri* associated with the *kcpA* locus and a 140-kilodalton protein. Infect Immun 57: 477–486

Payne SM, Finkelstein RA (1977) Detection and differentiation of iron-responsive avirulent mutants on Congo red agar. Infect Immun 18: 94–98

Sakai T, Sasakawa C, Makino S-I, Kamata K, Yoshikawa M (1986a) Molecular cloning of a genetic determinant for Congo red binding ability which is essential for the virulence of *Shigella flexneri*. Infect Immun 51: 476–482

Sakai T, Sasakawa C, Makino S-I, Yoshikawa M (1986b) DNA sequence and product analysis of the *virF* locus responsible for Congo red binding and cell invasion in *Shigella flexneri* 2a. Infect Immun 54: 395–402

Sakai T, Sasakawa C, Yoshikawa M (1988) Expression of four virulence antigens of *Shigella flexneri* is positively regulated at the transcriptional level by the 30 kilodalton *virF* protein. Mol Microbiol 2: 589–597

Sankaran K, Ramachandran V, Subrahamanyam YBK, Rajarathnam S, Elango S, Roy RK (1989) Congo red-mediated regulation of levels of *Shigella flexneri* 2a membrane proteins. Infect Immun 57: 2364–2371

Sansonetti PJ, Arondel J (1989) Construction and evaluation of a double mutant of *Shigella flexneri* as a candidate for oral vaccination against shigellosis. Vaccine 7: 443–450

Sansonetti PJ, Kopecko DJ, Formal SB (1981) *Shigella sonnei* plasmids: evidence that a large plasmid is necessary for virulence. Infect Immun 34: 75–83

Sansonetti PJ, Kopecko DJ, Formal SB (1982) Involvement of a plasmid in the invasive ability of *Shigella flexneri*. Infect Immun 35: 852–860

Sansonetti PJ, d'Hauteville H, Ecobichon C, Pourcel C (1983a) Molecular comparison of virulence plasmids in *Shigella* and enteroinvasive *Escherichia coli*. Ann Microbiol 134A: 295–318

Sansonetti PJ, Hale TL, Dammin GJ, Kapfer C, Collins Jr HH, Formal SB (1983b) Alterations in the pathogenicity of *Escherichia coli* k-12 after transfer of plasmid and chromosomal genes from *Shigella flexneri*. Infect Immun 39: 1392–1402

Sansonetti PJ, Ryter A, Clerc P, Maurelli AT. Mounier J (1986) Multiplication of *Shigella flexneri* within HeLa cells: lysis of the phagocytic vacuole and plasmid-mediated contact hemolysis. Infect Immun 51: 461–469

Sasakawa C, Yoshikawa M (1987) A series of Tn*5* variants with various drug-resistance markers and suicide vector for transposon mutagenesis. Gene 56: 283–288

Sasakawa C, Kamata K, Sakai T, Murayama SY, Makino S-I, Yoshikawa M (1986a) Molecular alteration of the 140-megadalton plasmid associated with loss of virulence and Congo red binding activity in *Shigella flexneri*. Infect Immun 51: 470–475

Sasakawa C, Makino S-I, Kamata K, Yoshikawa M (1986b) Isolation, characterization, and mapping of Tn*5* insertions into the 140-megadalton invasion plasmid defective in the mouse Sereny test in *Shigella flexneri* 2a. Infect Immun 54: 32–36

Sasakawa C, Kamata K, Sakai T, Makino S-I, Yamada M, Okada N, Yoshikawa M (1988) Virulence-associated genetic regions comprising 31 kilobases of the 230-kilobase plasmid in *Shigella flexneri* 2a. J Bacteriol 170: 2480–2484

Sasakawa C, Adler B, Tobe T, Okada N, Nagai S, Komatsu K, Yoshikawa M (1989) Functional organization and nucleotide sequence of virulence Region-2 on the large virulence plasmid of *Shigella flexneri* 2a. Mol Microbiol 3: 1191–1201

Savelkoul PH, Willshaw GA, McConell MM, Smith HR, Hamer AM, van der Zeijst BAM, Gasstra W (1990) Expression of CFA/I fimbriae is positively regulated. Microb Pathog 8: 91–99

Sereny B (1957) Experimental keratoconjunctivitis shigellosa. Acta Microbiol Hung 4: 367–376

Smith BR, Schleif R (1978) Nucleotide sequence of the L-arabinose regulatory region of *Escherichia coli*. J Biol Chem 253: 6931–6933

Stover CK, Vodkin MH, Oaks EV (1987) Use of conversion adaptors to clone antigen genes in λ gt11. Anal Biochem 163: 398–407

Tilney L, Portnoy DA (1989) Actin filaments and the growth, movement and spread of the intracellular bacterial parasite, *Listeria monocytogenes*. J Cell Biol 109: 1597–1608

Tobin JF, Schleif R (1987) Positive regulation of the *Escherichia coli* L-rhamnose operon is mediated by the products of tandemly repeated regulatory gene. J Mol Biol 196: 789–799

Venkatesan MM, Buysse JM (1990) Nucleotide sequence of invasion plasmid antigen gene *ipaA* of *Shigella flexneri* 5. Nucleic Acids Res 18: 1648

Venkatesan M, Buysse JM, Kopecko DJ (1988a) Characterization of invasion plasmid antigen (*ipaBCD*) genes from *Shigella flexneri*: DNA sequence analysis and control of gene expression. Proc Natl Acad Sci USA 85: 9317–9321

Venkatesan M, Buysse JM, Kopecko DJ (1988b) Development and testing of invasion-associated DNA probes for the detection of *Shigella* spps. and enteroinvasive *Escherichia coli*. J Clin Microbiol 26: 261–266

Venkatesan MM, Buysse JM, Kopecko DJ (1989) Use of *Shigella flexneri ipaC* and *ipaH* gene sequences for the general identification of *Shigella* spp. and entroinvasive *Escherichia coli*. J Clin Microbiol 27: 2687–2691

Watanabe H, Nakamura A (1985) Large plasmids associated with virulence in *Shigella* species have a common function necessary for epithelial cell penetration. Infect Immun 48: 260–262

Watanabe H, Nakamura A (1986) Identification of *Shigella sonnei* form I plasmid genes necessary for cell invasion and their conservation among *Shigella* species and entroinvasive *Escherichia coli*. Infect Immun 53: 352–358

Watanabe H, Arakawa K, Ito K, Kato J-I, Nakamura A (1990) Genetic analysis of an invasion region by use of a Tn*3-lac* transposon and identification of a second positive regulator gene, *invE*, for cell invasion of *Shigella sonnei*: significant homology of InvE with ParB of plasmid P1. J Bacteriol 172: 619–629

Yamada M, Sasakawa C, Okada N, Makino S-I, Yoshikawa M (1989) Molecular cloning and characterization of chromosomal virulence region *kcpA* of *Shigella flexneri*. Mol Microbiol 3: 207–213

Shigella Lipopolysaccharide: Structure, Genetics, and Vaccine Development

H. N. BRAHMBHATT[1], A. A. LINDBERG[2], and K. N. TIMMIS[1]

1 Introduction

Lipopolysaccharide (LPS) is a compound macromolecule anchored in the outer leaflet of the outer membrane of Gram-negative bacteria (Fig. 1) and extending out from the cell into the external medium (for detailed reviews see JANN and JANN 1984; MÄKELÄ and STOCKER 1984). It is a major structural component of the cell surface, and it has been calculated that there are about 2.5×10^6 molecules per cell in *Salmonella typhimurium*, occupying some 45% of the surface of the outer membrane (INOUYE 1979). Each LPS molecule is composed of three distinct structural segments, namely the innermost hydrophobic lipid A moiety, which constitutes the main lipid component of the outer leaflet of the asymmetrical outer membrane; the outermost O-specific polysaccharide (also called the O-antigen or somatic antigen), which consists of a short or

[1] Department of Microbiology, GBF, National Research Center for Biotechnology, Mascheroder Weg 1, 3300 Braunschweig, Federal Republic of Germany

[2] Department of Clinical Bacteriology, Karolinska Institutet, Huddinge University Hospital, Huddinge, Sweden

Current Topics in Microbiology and Immunology, Vol. 180

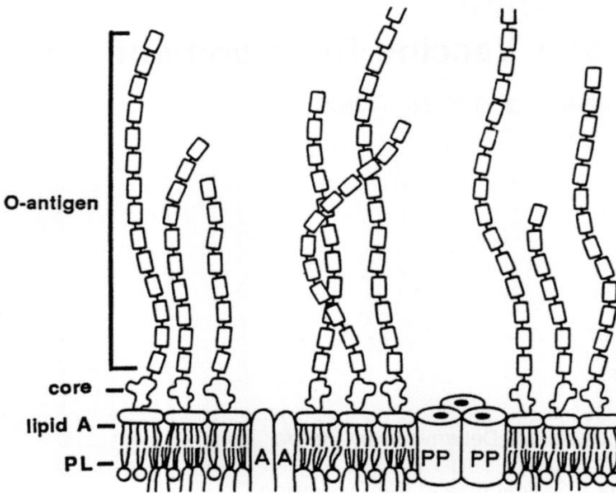

Fig. 1. Molecular architecture of the outer membrane of the cell envelope of enterobacteriaceae. *PL*, phospholipid, *PP*, pore protein; *A*, OmpA protein. (Adapted from MAAGD and LUGTENBERG 1987)

long linear polymer of an oligosaccharide repeat unit; and the core oligosaccharide, which links the O-antigen to the lipid A. LPS molecules interact structurally and functionally with a number of other cell surface components, including several outer membrane proteins (LUGTENBREG and VAN ALPHEN 1983; NIKAIDO and VAARA 1985).

In addition to its structural role, the O-polysaccharide fulfils a number of important functions, such as providing resistance to nonspecific, e.g., inducing activation and nonproductive deposition of complement far from its membrane target (BUCHANAN and PEARCE 1979; MÄKELÄ et al. 1980; MIMS 1982; PENN 1983; PLUSCHKE et al. 1983; FALKONE et al. 1984; GOLDMAN et al. 1984; JOINER et al. 1984, 1986; TIMMIS et al. 1985), and resistance to phagocytosis (MÄKELÄ et al. 1980; MIMS 1982) and immune defenses of animals, in the case of animal pathogens. The O-polysaccharide is also a major cell surface antigen and among different bacteria exhibits considerable structural and antigenic diversity, a feature that provides the basis of the serological classification of many Gram-negative bacteria (KAUFFMANN 1961, 1966). O-antigens are also receptors for many bacteriophages (LINDBERG 1973; HANNECART-POKORNI et al. 1976) and hence play a role in bacteriophage typing as well as host-parasite interactions in microbial ecology.

The core oligosaccharide, on the other hand, exhibits less structural diversity: only one core structure has thus far been identified in *Salmonella* (JANN and WESTPHAL 1975; LÜDERITZ et al. 1972), whereas five different core structures have been reported in *Escherichia coli* (JANSSON et al. 1981). Lipid A, whose structure is highly conserved in Gram-negative bacteria (for reviews see WOLLENWEBER and RIETSCHEL 1990; RAETZ 1990), is a potent activator of macrophages and causes the rapid induction and synthesis of tumor necrosis factor (mediator of endotoxic events) (BEUTLER and CERAMI 1988; KIENER et al. 1988), interleukin 1 (pyrogen) (LOPPNOW et al. 1989), and other proteins (WOLPE et al. 1988). It is responsible for a wide spectrum of pathophysiological host reactions, such as fever, hypotension, leukopenia followed by leukocytosis, disseminated intravascular coagulation, etc., some of which lead to irreversible shock with a fatal outcome that occurs particularly during disseminated infections (KADIS et al. 1971). Because of its toxicity, LPS is also termed endotoxin (WESTPHAL et al. 1983).

Much of our knowledge of LPS structure and biosynthesis comes from early elegant classical genetic and biochemical studies. Progress with these approaches, however, slowed as a result of the complexity of the metabolic processes, the difficulty of the biochemistry, the essential nature of some of the catalytic steps (housekeeping functions), and the confounding effects of undetected secondary mutations. The circumvention of some of these problems through the application of gene cloning, the precision of site-specific mutagenesis, and the use of newer biochemical methods promises to herald a second renaissance in LPS biology.

2 Structure of *Shigella* O-Antigens

O-antigens consist of linear polymers of repeat units composed of di- to hexasaccharides. Among shigellae many different serotypes have been recognized: 12 for *Shigella dysenteriae*, 14 for *Shigella flexneri*, 15 for *Shigella boydii*, and one for *Shigella sonnei* (for review see EWING and LINDBERG 1984). The structural basis of these antigenic specificities has been reviewed by KENNE and LINDBERG (1983). In *S. dysenteriae* each serotype has a unique repeat unit structure (Table 1; EWING and LINDBERG 1984; KENNE and LINDBERG 1983) with no significant cross-reactivities between serotypes. In contrast, all *S. flexneri* serotypes except six are composed of a common tetrasaccharide repeating unit (EWING and LINDBERG 1984): →2-α-L-Rhap1→2-α-L-Rhap1→3-α-L-Rhap1→3-β-D-GlcpNA1→whose structure exhibits serological specificity Y. Other serospecificities within the species result from α-D-glucopyranosyl and O-acetyl group substitutions on the basic repeating unit. In *S. boydii*, the structure of only serotype 6 has been characterized (Table 1; EWING and LINDBERG 1984). The O-antigen of *S. sonnei* is composed of a disaccharide repeat unit (Table 1; EWING and LINDBERG 1984) which is also shared by the LPS of *Plesiomonas shigelloides* (Table 1). Many cross-reactivities exist between certain *Shigella* serotypes and *E. coli* O-groups. For example, the O-polysaccharide of *S. dysenteriae* serotypes 2 and 3 are identical to *E. coli* serotypes O112 and O124, respectively. Such cross-reactions have also been found between *S. flexneri* and *E. coli* O-groups. Several cross-reactivities have been shown within *S. boydii*, between *S. boydii* and other shigellae and between *S. boydii* and other enterics, although the structural bases of these relationships have not been determined.

Table 1. O-specific polysaccharide chain structures in *S. dysenteriae* serotypes, *S. sonnei*, and *S. boydii* serotype 6

	Structure
S. dysenteriae serotype	
1	→3)-α-L-Rha*p*-(1→3)-α-L-Rha*p*-(1→2)-α-D-Gal*p*-(1→3)-α-D-Glc*p*NAc-(1→
2	→3)-α-D-Gal*p*NAc-(1→3)-α-D-Gal*p*NAc-(1→4)-α-D-Glc*p*-(1→4)-β-D-Gal*p*-(1→ ↑4 (on the second α-D-Gal*p*NAc) |1 AcO 3/4-α-D-Glc*p*NAc
3	→3)-β-D-Gal*p*NAc-(1→3)-α-D-Gal*p*-(1→6)-β-D-Gal*f*-(1→ ↑4 (on α-D-Gal*p*) |1 β-D-Glc*p*LcA-(1→6)-α-D-Gal*p*
4	→3)-α-D-Glc*p*NAc-(1→3)-α-D-Glc*p*NAc-(1→4)-α-D-Glc*p*A-(1→3)-α-L-Fuc*p*-(1→ ↑4 (on the second α-D-Glc*p*NAc) |1 AcO-α-L-Fuc*p*
5	→3)-β-D-Glc*p*NAc-(1→4)-α-D-Man*p*-(1→4)-α-D-Man*p*-(1→ ↑3 (on the first α-D-Man*p*); ↑2(3) (on the second α-D-Man*p*) |1 α-L-Rha*p*LcA; OAc
6	→3)-β-D-Gal*p*NAc-(1→3)-α-D-Gal*p*-(1→6)-α-D-Glc*p*-(1→ ↑4 (on α-D-Gal*p*) |1 X
8	→4)-β-D-Gal*p*A-(1→3)-β-D-Gal*p*NAc-(1→3)-β-D-Gal*p*NAc-(1→ ↑4 (on the second β-D-Gal*p*NAc) |1 β-D-Glc*p*NAc-(1→4)-β-D-Glc*p*
9	→4)-α-D-Gal*p*-(1→3)-β-D-Glc*p*NAc-(1→3)-β-D-Gal*p*-(1→4)-β-D-Man*p*-(1→ ↑2(3) (on α-D-Gal*p*): OAc 4, 6 (on β-D-Gal*p*): HO_2C / CH_3 (pyruvate acetal)
10	→2)-β-D-Man*p*-(1→3)-α-D-Man*p*NAc-(1→3)-β-L-Rha*p*-(1→4)-D-Glc*p*NAc-(1→
S. sonnei	→4)-AltNAcUA-(1→3)-α-4-NH_2-FucNAc-(1→
S. boydii 6	→3)-GlcNAc-(1→3)-β-Gal-(1→6)-α-Man- ↑ (on β-Gal) GlcUA

AcO, O-acetyl group; D-Gal, D-galactose; D-GalNAc, *N*-acetyl-D-galactosamine; D-Glc, D-glucose; D-GlcA, D-glucuronic acid; D-GlcLcA, 4-*O*-[(*R*)-1-carboxyethyl]-D-glucose; D-GlcNAc, *N*-acetyl-D-glucosamine; D-Man, D-mannose; D-ManNAc, *N*-acetyl-D-mannosamine; L-Fuc, L-fucose; L-Rha, L-rhamnose; L-RhaLcA, 3-*O*-[(*R*)-1-carboxyethyl]-L-rhamnose; AltNAcUA, 2-deoxy-2-acetamidolatrose; 4-NH_2-FucNAc, 2-acetamide-4-amino-2,4,6-trideoxy-D-galactopyranose; *p* and *f* indicate if the sugars are in the pyranose or furanose forms, respectively; X, unidentified component.

3 Genetics

3.1 LPS Core and O-Antigen Biosynthesis Are Determined by the *rfa* and *rfb* Gene Clusters

In *S. typhimurium* the O-antigen biosynthesis and assembly genes are encoded by the *his*-linked *rfb* gene cluster (44′ on the chromosome) and the *rfc* gene (approximate map position 32), while the enzymes involved in core polysaccharide biosynthesis are encoded by the *rfa* locus (located between *cysE* and *pyrE* at 81′ on the linkage map). A number of housekeeping genes localized at various other parts of the chromosome specify factors needed for the synthesis of constituents of both the core and O-polysaccharide. The *rfb*-encoded enzymes are, on the one hand, nucleotide diphosphate sugar synthetases and, on the other, glycosyl transferases which synthesize activated sugar derivatives and transfer the sugar moieties to the antigen carrier lipid (ACL; C-55 polyisoprenoid alcohol phosphate; ROBBINS and WRIGHT 1971), respectively, to build up the O-antigen repeat unit. Polymerization of repeat units, which also takes place on the ACL, is carried out by O-antigen polymerase (ROBBINS et al. 1966) encoded by the *rfc* locus. Ligation of the lipid-linked, polymerized O-polysaccharide to the lipid A-core is carried out by two enzymes encoded by *rfbT* and *rfaL* genes (SUBBAIAH and STOCKER 1964; GEMSKI and STOCKER 1967). Recent data from pulse-chase experiments show that polymerization of O-specific polysaccharide chains as well as attachment of O-antigen to core LPS takes place on the periplasmic face of the inner membrane (MCGRATH and OSBORN 1991). The completed LPS molecule is then transported to the outer leaflet of the outer membrane (for review see RAETZ 1990) where it becomes complexed with proteins and membrane phospholipids (reviewed by LUGTENBERG and VAN ALPHEN 1983).

The mechanism of O-polysaccharide biosynthesis is not the same in all bacteria. O-antigens composed of repeating units containing different sugars, for example, those of *Salmonella* groups A, B, D, and E, are assembled sequentially through the polymerization of the repeat units on the ACL. On the other hand, O-antigens, such as those of *Salmonella* groups C1 and L and of *E. coli* O8 and O9, consisting of mannose homopolymers are synthesized continuously through the transfer of individual sugar moieties to the growing chain on the α-glucosyl diphosphoundecaprenol carrier molecule (α-glucosyl-ACL). This mechanism of O-antigen biosynthesis is termed *rfe* dependent since the *rfe* gene product(s) (involved in enterobacterial common antigen biosynthesis) are necessary for the α-glucosylation of ACL, whereas O-polysaccharide biosynthesis in *Salmonella* groups A, B, D, and E is classified *rfe* independent. In contrast to the situation for O-antigen biosynthesis, the mechanism of core biosynthesis is the same in all bacteria studied so far, and the constituent sugars are sequentially transferred from their respective nucleotide diphosphate derivatives to the growing core oligosaccharide.

The backbone LPS structure can be modified with a variety of substitutions, often in a non-stoichiometric manner. Such modifications are determined by genetic loci outside the LPS gene clusters and in some cases located on genomes of temperate phages (for reviews see JANN and WESTPHAL 1975; LÜDERITZ et al. 1971).

Early genetic analyses of O-polysaccharide biosynthesis in *Shigella* mostly concentrated on *S. flexneri* serotype 2a. In this strain the O-polysaccharide biosynthesis genes were found in two distinct regions of the chromosome (FORMÁL et al. 1970). The "a" or group antigen corresponding to the O-repeat unit backbone is encoded by the *his*-linked *rfb* gene cluster, whereas the type antigen locus, which determines the "type 2" modification of the O-polysaccharide, is linked to the *pro* gene cluster in the vicinity of 6′ on the genetic map. The synthesis and assembly of core sugars in *S. flexneri* is encoded by the *rfa* locus (GODARD and HANNECART-POKORNI 1977). Despite the early emphasis on O-antigen genetics of *S. flexneri* 2a, recent studies using molecular genetic approaches facilitated by the development of Western blotting procedures for electrophoretically fractionated LPS (NURMINEN et al. 1984; SEID et al. 1984; STURM et al. 1984) have tended to concentrate on the O-antigen of Shiga's bacillus, *S. dysenteriae* serotype 1. These studies have revealed that both plasmid (a 9-kb plasmid originally designated pHW400; WATANABE and TIMMIS 1984; WATANABE et al. 1984) and chromosomal genes (HALE et al. 1984; STURM and TIMMIS 1986) are required for O-antigen production in this organism. Transfer of the *his*-linked *rfb* chromosomal locus from *S. dysenteriae* 1 into *E. coli* K-12 carrying the 9-kb plasmid of *S. dysenteriae* 1 resulted in the synthesis of *S. dysenteriae* 1 O-specific LPS in K-12 (HALE et al. 1984; STURM and TIMMIS 1986). The plasmid-carried determinant was designated *rfp* (WATANABE et al. 1984). In vivo cloning with RP4::miniMu facilitated the isolation and characterization of the *rfb* locus from the chromosome of *S. dysenteriae* 1 and its fusion to a 4.7-kb DNA fragment carrying the *rfb* plasmid determinant to obtain an *rfb-rfp* gene cassette which, when transformed into *E. coli* K-12, directs the synthesis of *S. dysenteriae* 1 O-specific LPS (STURM and TIMMIS 1986; STURM et al. 1986a, b). Insertion and deletion mutagenesis of the *rfb* and *rfp* gene clusters, followed by analysis of the sensitivity/resistance of host bacteria to "rough-specific" bacteriophages such as T3 and of mutant LPS phenotypes on sodium dodecyl sulfate–polyacrylamide gel electrophoresis (SDS-PAGE) and reactivity with anti-O antibody in immunoblots permitted the construction of a tentative physical and functional map of the gene clusters (STURM et al. 1986a, b). Sequencing of these genes and detailed chemical composition analysis of LPS purified from various mutant bacteria is in progress. Limitations of a genetic analysis of the O-antigen of *S. sonnei* was also recently reported (YOSHIDA et al. 1991). An interesting aspect of the PAGE analysis of LPS prepared from mutant bacteria is that new electrophoretic species that may correspond to O-antigen biosynthesis intermediates were detected (STURM et al. 1986b; MILLS and TIMMIS 1988). The possibility of correlating particular genetic defects with specific electrophoretic species on SDS gels that can subsequently be chemically and immunochemically characterized would greatly simplify the genetic analysis of O-antigen biosynthesis.

3.2 Extrachromosomal Elements Play a Critical Role in the Expression of O-Antigens and in Their Serospecificities

In recent years, a number of O-antigens have been shown to be partly or completely encoded by plasmid-borne determinants. Organisms having such O-antigens include *S. sonnei* (KOPECKO et al. 1980; HALE et al. 1983), *S. dysenteriae* 1 (WATANABE and TIMMIS 1984; WATANABE et al. 1984), some *Salmonella* serotypes (POPOFF and LE MINOR 1985), and an enteropathogenic strain of *E. coli* (EPEC) serotype O111:NM (RILEY et al. 1987, 1990). *S. sonnei* strains when grown in laboratory culture spontaneously yield "rough" avirulent variants that exhibit a complete lipid A-core, but lack the O-polysaccharide (ROMANOWSKA and REINHOLD 1973; GAMIAN and ROMANOWSKA 1982). The instability of O-antigen production in this organism was found to be due to the fact that O-antigen biosynthesis is entirely encoded by a 120-MDa nonconjugative virulence plasmid (KOPECKO et al. 1980; SANSONETTI et al. 1980, 1981b; BINNS et al. 1985), which in culture is spontaneously lost at high frequency and thereby causes phase variation.

In *S. dysenteriae* 1 strain W30864 it was shown that a 9-kb plasmid, pHW400, encodes one or more functions for O-polysaccharide synthesis and for bacterial virulence (WATANABE and TIMMIS 1984). Gene cloning and Tn5 mutagenesis experiments permitted localization of the essential locus involved in O-antigen biosynthesis, designated *rfp*, to a 2-kb region of the plasmid. Southern hybridization of an *rfp* gene-containing DNA sequence to a variety of shigellae, salmonellae, and *E. coli* strains revealed that the *rfp* genes are present in other *S. dysenteriae* 1 strains, also on 9-kb pHW400-like plasmids, but not in other serogroups of *Shigella* or other species examined.

The O-antigen repeat unit has the structure -glcNAc-rha-rha-gal-. *S. dysenteriae* 1 derivatives lacking pHW400 do not produce O-antigen. *E. coli* K-12 containing either the *rfb* region of *S. dysenteriae* 1 or the *rfp* locus also fails to produce *S. dysenteriae* 1-specific O-antigen. Introduction of the *rfp* locus into pHW400-negative derivatives of *S. dysenteriae* 1 or *E. coli* K-12 already carrying the *S. dysenteriae* 1 *rfb* locus provokes O-antigen biosynthesis. Lowering the copy number of the *rfp* locus by transferring it from medium to high copy number vectors into the monocopy bacterial chromosome significantly reduces O-antigen expression in *E. coli* K-12 and *Salmonella* (S.D. MILLS et al., in preparation). Introduction of the *rfp* locus into *E. coli* K-12 on the medium copy number vector pACYC184, followed by SDS-PAGE fractionation of LPS and analysis by silver staining showed that low molecular weight species additional to the K-12 lipid A-core band were present. Chemical analysis of LPS prepared from such bacteria demonstrated the presence of an additional galactose residue linked to the terminal *N*-acetylglucosamine residue on the K-12 core, although galactose modification was incomplete and 30% of the core remained unmodified (STURM et al. 1986a; A.A. LINDBERG et al., in preparation). Although our understanding of O-antigen biosynthesis in *S. dysenteriae* 1 is still very incomplete, available data do suggest that the plasmid-specified function is a biochemically strategic, rate-limiting step in O-antigen production.

Temperate phages also play an important role in determining serotype specificities of *S. flexneri* strains (MATSUI 1958; ISEKI and HAMANO 1959; SIMMONS 1971; GEMSKI et al. 1975). Lysogenization by some of these phages is accompanied by integration of their DNA at a specific site, designated the Type or T locus (TAMAKOV et al. 1970), which is linked to the pro (6′) and *lac* (8′) loci in the *Shigella* chromosome, and by addition of glucose to the O-polysaccharide. Other converting phages cause addition of acetyl substituents to the backbone (SIMMONS 1971). These modifications are immunodominant and therefore produce new serotypes: for example, lysogenization of *S. flexneri* serotype Y strains (—:3,4) with phage SF6 results in O-acetylation of the O-antigen backbone and conversion to type 3b carrying group antigens 6,3 and 6,4 (GEMSKI et al. 1975; LINDBERG et al. 1972).

3.3 Transfer of O-Antigen Determinants to Heterologous Hosts Reveals O-Antigen-Core Linkage Specificities

O-antigen biosynthesis genes from *Shigella* have been transferred to and expressed in various heterologous hosts. The *rfb* genes of *S. flexneri* serotype 2a present on an in vivo constructed hybrid F′*lac*::miniMu plasmid and of *S. sonnei* present on the 120-MDa invasion plasmid of this organism have, for example, been transferred into *E. coli* K-12 and *S. typhi* Ty21a (FORMAL et al. 1981, 1984; BARON et al. 1987). Both O-polysaccharide species were expressed in these strains and covalently linked to the lipid A-core of K-12, but not to that of Ty21a (SEID et al. 1984; FORMAL et al. 1984; BARON et al. 1987). It appears that the unlinked heterologous O-polysaccharide was exported to the cell surface where it remained loosely attached in a form resembling capsular polysaccharide. Transfer of the *rfb-rfp* gene clusters of *S. dysenteriae* 1 into *E. coli* K-12, *aro* A mutant strains of *Salmonella typhimurium*, *Salmonella dublin*, and *S. typhi* Ty21a (STURM and TIMMIS 1986; MILLIS et al. 1988) also resulted in expression of both homologous and heterologous LPS species in these hybrid strains, as detected by agglutination of whole bacteria and immunoblot analysis of purified LPS. All strains except *S. typhi* Ty21a exhibited covalent linkage of the *S. dysenteriae* 1 O-polysaccharide to the lipid A-core structure of the host strains. Western blot analysis of the LPS from the hybrid Ty21a strain revealed the *S. dysenteriae* 1 O-polysaccharide as a faint smear in the upper parts of the gel, in contrast to that isolated from the other bacteria which appeared as a typical LPS ladder. Similar observations have been made with the *Vibrio cholerae* O1 serotype Inaba O-antigen (ATTRIDGE et al. 1990; TACKET et al. 1990).

Transfer of the *E. coli* K-12 *rfa* locus, which encodes core biosynthesis functions, to Ty21a produced a Ty21a hybrid strain that expressed both the *V. cholerae* and the *S. typhi* O-polysaccharides covalently linked to the lipid A-core.

These results show that the linkage of O-polysaccharides to the cores of *E. coli* K-12, *S. typhimurium,* and *S. dublin* has a relaxed specificity since structurally distinct O-polysaccharides from a variety of strains become linked to structurally distinct LPS cores of different heterologous hosts. This relaxedness does not, however, extend to *S. typhi* Ty21a. It is not clear at this point whether the nonpermissiveness of the Ty21a core for heterologous O-antigens reflects a greater specificity in the ligation reaction in *S. typhi* or whether it is a particular characteristic of the Ty21a strain which was made by nitrosoguanidine mutagenesis and which may have accumulated other mutations that affect the linkage. Current work on the genetic characterization of the *rfa* locus of *E. coli* K-12 (see, for example, AUSTIN et al. 1990; PEGUES et al. 1990) will facilitate elucidation of the molecular basis of these specificities and delineation of the functions required to link heterologous O-polysaccharides to the lipid A-core structures of host strains such as Ty21a.

4 Vaccine Development

Bacillary dysentery is an invasive disease of the colon in humans and higher primates and is transmitted via the fecal-oral route. It is highly infectious and as few as ten bacteria can produce disease in a healthy human. Dysentery is a classical disease associated with poor hygiene, overcrowding, and stress. The majority of cases thus occur in regions of developing countries having suboptimal sanitation, but outbreaks have also been associated with war zones and mental institutions in developed countries. The causative agents are various species of *Shigella* and enteroinvasive strain of *E. coli* (EIEC) (GEMSKI and FORMAL 1975) present in endemic areas, while the most severe form of the disease is caused by *S. dysenteriae* 1 (KEUSCH and BENNISH 1988). The pathology of the disease includes local destruction of the colonic mucosa and the formation of abscesses as a result of an acute inflammatory response in the invaded areas. *S. dysenteriae* 1 infections in particular may also lead to serious complications such as hemolytic uremic syndrome (HUS), hypoproteinemia, leukemoid reactions, and sepsis (LEVINE 1982). According to a recent estimate there are more than 100 million cases of bacillary dysentery world-wide and approximately 600 000 deaths per year. Major problems in the clinical management of the disease are that it is not amenable to oral rehydration therapy, and most *Shigella* isolates are resistant to a variety of antibiotics, particularly those generally available in developing countries (FARRAR 1985). The magnitude and severity of shigellosis and the difficulties associated with its treatment are the reasons why the development of efficacious *Shigella* vaccines are an urgent priority.

In early studies parenteral vaccines consisting of either killed shigellae or attenuated live bacteria were found not to be protective (FORMAL et al. 1967; reviewed by FORMAL and LEVINE 1984). It is widely assumed that, since shigellosis is a localized infection of the mucosal and submucosal layers of the intestine, it may not be susceptible to humoral immunity (see, for example, HIGGINS et al. 1955; DEVINO 1959). On the other hand, epidemiological data show that individuals having once suffered an episode of shigellosis are protected from reinfection by strains of the same serotype (DUPONT et al. 1972). This suggests that protective immunity will be attainable and that live vaccines which induce mucosal serotype-specific immunity, i.e., to bacterial O-antigen, may be most effective. Several studies using live, attenuated, orally administered *Shigella* vaccines have shown that they can be effective in eliciting protection in both experimental animals and humans (FORMAL et al. 1965, 1966a, b; MEL et al. 1965a, b, 1971; DUPONT et al. 1972; KETYL et al. 1974). However, most strains were found to revert to the virulant state and thus to be unsafe for human use. With the availability of more precise genetic tools that permit the construction of attenuated strains carrying defined deletion mutations that do not revert, the problem of re-acquisition of virulence is considerably diminished. Vaccine candidates that are currently favored by a number of groups are avirulent live organisms which, when administered orally, colonize the intestinal mucosa and stimulate a local immune response to the *Shigella* O-antigen. The precise mechanisms of protection in previously infected individuals are not clearly understood, but it is assumed that efficacious vaccines will need to stimulate both a secretory immunoglobulin A (IgA) response against the bacterial surface O-antigens as well as a cell-mediated immune response against infected host cells.

A number of strategies have been and are being employed to generate *Shigella* vaccines. Three candidate vaccines have been developed by Formal and colleagues, one directed against *S. flexneri* 2a, the second against *S. sonnei*, and the third is a bivalent vaccine against *S. typhi* and *S. flexneri* 2a infections. The first candidate vaccine employs as a delivery system an *E. coli* K-12 strain harboring the *Shigella* invasion plasmid and also carries the genes for the *S. flexneri* 2a somatic antigen (SANSONETTI et al. 1983; FORMAL et al. 1984). These bacteria penetrate the intestinal epithelium of ligated rabbit ileal loops and produce a mild inflammatory response, but are avirulent. When administered orally in rhesus monkeys, this strain conferred significant protection against oral challenge with virulent *S. flexneri* 2a. However, when administered in humans, a similar strain elicited diarrhea in 20% of the volunteers (HALE and FORMAL 1990). This problem has been partly overcome by the introduction of an *aroD* mutation into the *E. coli* carrier strain, although testing of the new vaccine in human volunteers revealed that a proportion of vaccinees still experienced intestinal cramps (KOTLOFF et al., unpublished observation; HALE 1990).

The second and third vaccines consist of *S. typhi* Ty21a as a delivery strain harboring the large plasmid of *S. sonnei* (specifying the O-antigen) and the

S. flexneri 2a O-antigen biosynthesis genes, respectively (FORMAL et al. 1981; BARON et al. 1987). However, in neither hybrid strain does the heterologous O-antigen link to the lipid A-core of Ty21a (see above), and only very weak immune responses were obtained against this unlinked O-polysaccharide, presumably because it exists as a hapten.

An alternative approach is to introduce nonreverting attenuating mutations into virulent *Shigella* strains. These strains would have the advantage of being able to stimulate local immune responses to a multitude of surface antigens of the virulent bacteria (reviewed by HALE and FORMAL 1989). A recent promising candidate vaccine is a virulent *S. flexneri* serotype Y strain made auxotrophic for aromatic metabolites, including *p*-aminobenzoic and (A.A. LINDBERG et al., submitted), which are not available in mammalian tissues. Other vaccine candidates carrying attenuating mutations such as *aroA*, *aroD*, *icsA* and others have also been recently described (LINDBERG et al. 1988; SANSONETTI and ARONDEL 1989; FONTAINE et al. 1990; LINDE et al. 1990).

Hybrid antigen delivery strains, namely *S. typhimurium* and *S. dublin aroA* mutants, and *S. typhi* Ty21a carrying a plasmid-borne *rfb-rfp* gene cassette encoding the *S. dysenteriae* 1 O-antigen have also been constructed (MILLS et al. 1988). In laboratory culture the hybrid salmonellae stably express the *Shigella* O-antigen, but only when antibiotic selection pressure for the plasmid is maintained. In order to stabilize heterologous O-antigen expression in these strains, the *rfb-rfp* gene cassette was integrated into the chromosome of the *S. typhimurium* and *S. dublin aroA* mutant via a mercury resistance transposon (S.D. MILLS et al., in preparation). These hybrid strains stably express both the homologous and the heterologous LPS species and are currently being tested in animal models for their ability to stimulate protective immunity to *S.* dysenteriae 1 infection.

One of the problems facing the development of live oral bacterial vaccines is the current requirement of the Food and Drug Administration in the United States (FDA) and other authorizing agencies that such vaccines do not carry resistances to antibiotics currently in clinical use. This problem and that of vector plasmid instability were recently resolved with the development of a range of versatile transposon cloning-vectors containing nonantibiotic resistance selection markers for the stable chromosomal integration of cloned genes (HERRERO et al. 1990).

5. Role of the O-Polysaccharide in Bacterial Virulence

Since rough derivatives (devoid of O-antigen) of many Gram-negative bacterial pathogens are avirulent and concomitantly exhibit susceptibility to serum killing, opsonization, and phagocytosis, the O-antigen is thought to contribute to virulence by mediating resistance to host defense mechanisms. Rough shigellae are also avirulent (OKAMURA and NAKAYA 1977; OKAMURA et al. 1983; SANSONETTI et al. 1981a, b), and, in experiments to convert *E. coli* into a pathogenic bacterium by the introduction of *S. flexneri* genes, O-antigen synthesis was found to be an essential component of the virulence factors required for the conversion (SANSONETTI et al. 1981a, b).

Early studies on the role of the *Shigella* O-antigen in virulence involved the use of spontaneous rough mutants, whose mutations were undefined, but probably resulted from large DNA deletions that may have inactivated genes other than those of O-antigens (for more detailed discussion of this point, see TIMMIS et al. 1986b) and of O-antigen determinants transferred on large, undefined DNA segments that could well have carried other virulence determinants. Causal relationships between O-antigens and virulence could not therefore be established with such experiments. In order to generate O-antigen-specific mutations, BINNS et al. (1985) developed a convenient transposon mutagenesis system and used it to isolate O-antigen-negative Tn*3* and Tn*5* mutants of *S. sonnei* and *S. dysenteriae* 1 (Table 2). All O-minus mutants were avirulent and failed to provoke a positive Sereny reaction, but retained the ability to invade HeLa cells (though a reduction in the invasive ability of *S. dysenteriae* 1 mutants was noted). These results demonstrated a

Table 2. Virulence and phage sensitivity of O-antigen-negative mutants of *S. sonnei* (F3905/81) and *S. dysenteriae* 1 (W30864). (Data from BINNS et al. 1985)

Strain	O-antigen	HeLa cell invasion	Sereny test	Bacteriophages				
				Fo	Fp1	C21	6SR	BR10
S. sonnei								
Form 1	S	+	+	−	−	−	−	−
Form 1::Tn*5*	S	+	+	−	+	−	+	+
Form 1::Tn*5*::Tn*3*-1	R	+	−	−	+	−	+	+
Form 1::Tn*5*::Tn*3*-2	R	+	−	−	+	−	+	+
Form 1::Tn*5*::Tn*3*-3	R	+	−	−	+	−	+	+
Form II	R	−	−	−	+	−	+	+
S. dysenteriae 1								
W30864	S	+	+	−	−	−	−	−
W30864 + pMB500	S	+	+	NT	NT	NT	NT	NT
W30864::Tn*5*-2F4	R	+	−	−	+	−	+	+
W30864::Tn*5*::51A8	R	+	−	−	+	−	+	+
W30864::Tn*5*-93E12	R	+	−	−	+	−	+	+
W30864::Tn*5*-93E12 + pHW401	S	+	+	−	+	−	+	+

Smooth (S) and rough (R) were determined with specific antiserum. NT, not tested. In *S. sonnei*, transposon Tn*5* was used to tag and stabilize the form 1 plasmid and transposon Tn*3* was used to generate O-antigen-negative mutants (mutants designated Tn*3*-1, Tn*3*-2 and Tn*3*-3). Plasmid pMB500 was used as a transposon Tn*5* donor, and the *S. dysenteriae* 1 O-antigen-negative mutants generated are designated Tn*5*-2F4, Tn*5*::51A8, and Tn*5*-93E12. Plasmid pHW401 is a transposon Tn*801*-tagged derivative of the 9-kb plasmid (carrying the *rfp* determinant; WATANABE and TIMMIS 1984) causal relationship between O-antigen production and bacterial virulence in *S. sonnei* and *S. dysenteriae* 1.

While the O-polysaccharide is not required for penetration of epithelial cells, it has been proposed to play a role in bacterial colonization of the intestinal mucosa (IZHAR et al. 1982). This suggestion remains, however, to be confirmed. In a recent interesting study (OKADA et al. 1991), mutants of *S. flexneri* were isolated which had LPS characteristics such as (a) altered core structure; (b) absence of O-polysaccharide; (c) decreased length of O-polysaccharide; and (d) absence of core structure but with intact O-polysaccharide. The first three types of mutants were found to be capable of invading MK2 cells, intracellular multiplication, and spread within infected cells (within 8 h of infection), but to be incapable of intercellular spread. Four hours postinfection, mutant bacteria showed a tendency to swell and be rough. These findings suggest that the O-antigen of shigellae is needed for intercellular spread and for protecting intracellular bacteria from damage induced by epithelial cell defense mechanisms, in addition to any role it may have in protecting extracellular bacteria from host defenses.

Several reports suggest that the chemical composition and structure of O-antigens may influence the virulence of the bacteria. Evidence from analyses of the virulence of salmonellae has shown that the more virulent Salmonella serotype BO bacteria (like *S. typhimurium*) do not activate complement via the alternative pathway as efficiently as serogroup DO and CO bacteria, and show a higher resistance to phagocytosis. The molecular basis of the different complement—activating activities was shown using saccharide-lipid conjugates to be a consequence of the structure of the O-antigen repeat unit. *S. dysenteriae* serotype 1 bacteria is more virulent than serotypes 2–10 and other *Shigella* species. The possibility that structural differences in *Shigella* O-antigens may also be important for the virulence differences among *Shigella* species deserves further investigation.

6 Conclusion

LPS constitutes a major structural and antigenic component of the surface of Gram-negative bacteria and plays a number of important functional roles, including that of a scaffold for other cell surface components and providing partial resistance to a number of noxious environmental agents, such as complement and phagocytes, in the case of bacteria whose niches are on/in the animal body. Because LPS is a major cell surface component, is a virulence factor of many Gram-negative pathogens, and is highly immunogenic, it is an important immunogen of many vaccines. However, the complexity of its structure, biochemistry, and genetics (which result mostly from the fact that it is a nonprotein heteropolymer) causes its precise engineering, both into live vaccine strains and into safe, convenient LPS production strains to produce new generation vaccines to be a considerable technical challenge. Rapidly accumulating knowledge on the specificities of core-O-antigen ligation reactions and on the molecular genetics of the core and O-antigens (see, for example, BRAHMBHATT et al. 1986, 1988; VERMA et al. 1988, VERMA and REEVES 1989; WYK and REEVES 1989; AUSTIN et al. 1990; PEGUES et al. 1990; JIANG et al. 1991) will, however, considerably accelerate progress in this area.

Rapid advances in the molecular genetics of LPS biosynthesis will both facilitate and be facilitated by progress in LPS biochemistry. This, in turn, will lead to elucidation of structure:function relationships in LPS, characterization of critical interactions of LPS with other cellular components, elucidation of molecular mechanisms of LPS activities crucial to microbial pathogenesis, and a detailed description of the interrelationships of biosynthetic reactions that are in part specific for LPS and in part shared with other cellular components (e.g., enterobacterial common antigen, capsular polysaccharides). As a result, potential LPS-specific targets for new chemotherapeutic agents may be identified.

One of the surprising features of LPS genetics is the involvement of extrachromosomal elements in the biosynthesis of O-antigens, which are generally regarded as major structural components of the cell that mediate functions important for bacterial survival in hostile environments and would therefore be expected to be chromosomally encoded like other crucial cell components. Why, for example, the entire O-antigen determinant of *S. sonnei* is located on an unstable plasmid that is spontaneously lost at high frequency in culture (and presumably in other conditions in which the bacterium is not in its pathogenic growth mode) is unclear. Perhaps O-antigen biosynthesis in this organism represents a particularly heavy physiological burden, and its facile loss during growth outside the animal host constitutes a selective advantage. The high infectivity of shigellae, which permits only a few O-antigen-positive organisms to initiate an infection in a new host, possibly compensates for O-antigen instability and assures propagation of the pathogenic population. Even more surprising, however, is the distribution of the O-antigen determinants of *S. dysenteriae* 1 on two separate genetic elements, namely the chromosome (the *rfb* gene cluster) and a 9-kb plasmid (the *rfp* determinant). In this bacterium, the 9-kb plasmid is reasonably stable so it is unlikely that there are similar causes of plasmid involvement in O-antigen biosynthesis in *S. sonnei* and *S. dysenteriae* 1. Recent studies indicate that the *rfp*-specified function, i.e., addition of a galactose moiety to the core oligosaccharide, is the first and rate-limiting step in O-antigen biosynthesis in *S. dysenteriae* 1; transfer of *rfp* to a chromosomal location diminishes O-antigen production. In this case, therefore, the presence of a key O-antigen gene on a stable high copy number plasmid may reflect the need for expression of the *rfp* gene at a higher level than that of the *rfb* gene cluster. Although gene dosage is considered to be a "primitive" mechanism for varying gene expression, it may be that this represents an early stage in a continuing evolutionary process towards a more sophisticated mechanism of differential expression of O-antigen determinants in this bacterium. In this regard, it should be emphasized that plasmids, as genetic elements that are nonessential for normal growth and survival of the cell and that can be mutated and lost and regained without major consequences for central metabolism and normal cellular activities, are generally considered to be important instruments of genetic change and evolution which permit bacterial populations to adapt to special environments (e.g., growth within epithelial cells which is rendered possible by the invasion genes of the virulence plasmid of shigellae and EIEC) and to rapidly evolve new activities through mutation, gene duplication and rearrangement, and gene transfer (see, for example, TIMMIS et al. 1986a). Phages that lysogenize shigellae can also influence the structure and serospecificity of O-antigens by causing modification (glucosylation or O-acetylation) of the repeat backbone. Whether or not this constitutes a selective advantage for the bacterium or the bacterium-phage interaction, and if so which one, is not at present clear, though antigenic variation in microbial populations is an important mechanism of circumventing individual and herd immunity.

Acknowledgements. Work from the GBF was supported by the Deutsche Forschunggemeinschaft (SFB1292 "Gastrointestinale Barriere"), and the work conducted by the Swedish group was supported by the Swedish Medical Research Council (grant No. 16X-656).

References

Attridge SR, Daniels D, Morona JK, Morona R (1990) Surface co-expression of *Vibrio cholerae* and *Salmonella typhi* O-antigens on Ty21a clone EX210. Microb Pathoge 8: 177–188

Austin EA, Graves JF, Hite LA, Parker CT, Schnaitman CA (1990) Genetic analysis of lipopolysaccharide core biosynthesis by *Escherichia coli* K-12: Insertion mutagenesis of the *rfa* locus. J Bacteriol 172: 5312–5325

Baron LS, Kopecko DJ, Formal SB, Seid R, Guerry P, Powell C (1987) Introduction of *Shigella flexneri* 2a Type and group antigen genes into oral typhoid vaccine strain *Salmonella typhi* Ty21a. Infect Immun 55: 2797–2801

Beutler B, Cerami A (1988) The history, properties, and biological effects of Cachetin. Biochemistry 27: 7575–7582

Binns MM, Vaughan S, Timmis KN (1985) 'O'-antigens are essential virulence factors of *Shigella sonnei* and *Shigella dysenteriae*. Zentralbl Balzteriol Miknrbiol Hyg [B] 181: 197–205

Brahmbhatt HN, Quigley NB, Reeves PR (1986) Cloning part of the region encoding biosynthetic enzymes for surface (O-antigen) of *Salmonella typhimurium*. Mol Gen Genet 203: 172–176

Brahmbhatt HN, Wyk PJ, Quigley NB, Reeves PR (1988) A complete physical map of the *rfb* locus encoding biosynthetic enzymes for the O-antigen of *Salmonella typhimurium* LT2. J Bacteriol 170: 98–102

Buchanan TM, Pearce WA (1979) Pathogenic aspects of outer membrane components of Gram-negative bacteria. In: Inouye M (ed) Bacterial outer membranes. Wiley, New York, pp 475–514

Charles I, Dougan G (1990) Gene expression and the development of live enteric vaccines. Trends Biochem 8: 117–121

Devino LV (1959) The specific prophylaxis of dysentery with vaccines from complete antigens. Zh Miobiol Epidemiol Immunobiol 30: 22–31

Dupont HL, Hornick RB, Snyder MJ, Libonati JL, Formal SB, Gangarosa EJ (1972) Immunity in shigellosis. II. Protection induced by oral live vaccine or primary infection. J Infect Dis 125: 12–16

Ewing WH, Lindberg AA (1984) Serology of *Shigella*. In: Bergan T (ed) Methods in Microbiology, vol 14. Academic, London, pp 113–142

Falkone G, Campa M, Smith H, Scott GM (1984) Bacterial and viral inhibition and modulation of host defences. Academic, London

Farrar WE (1985) Antibiotic resistance in developing countries. J Infect Dis 152: 1103–1107

Fontaine A, Arondel J, Sansonetti PJ (1990) Construction and evaluation of live attenuated vaccine strains of *Shigella flexneri* and *Shigella dysenteriae* 1. Res Microbiol 141: 907–912

Formal SB, Levine MM (1984) Shigellosis. In: Germanier R (ed) Bacterial vaccines. Academic, New York, pp 167–186

Formal SB, LaBrec EH, Palmer A, Falkow S (1965) Protection of monkeys against experimental shigellosis with attenuated vaccines. J Bacteriol 90: 63–68

Formal SB, Kent TH, Austin S, LaBrec EH (1966a) Fluorescent-antibody and histological study of vaccinated and control monkeys challenged with *Shigella flexneri*. J Bacteriol 91: 2368–2376

Formal SB, Kent TH, May HC, Palmer A, Falkow S, LaBrec EH (1966b) Protection of monkeys against experimental shigellosis with a living attenuated oral polyvalent dysentery vaccine. J Bacteriol 92: 17–22

Formal SB, Maenza RM, Austin S, LaBrec EH (1967) Failure of parenteral vaccines to protect monkeys against experimental shigellosis. Proc Soc Exp Biol Med 25: 347–349

Formal SB, Gemski P, Baron LS, LaBrec EH (1970) Genetic transfer of *Shigella flexneri* 2a antigens to *Escherichia coli* K-12. Infect Immun 1: 279–287

Formal SB, Baron LS, Kopecko DJ, Washington O, Powell C, Life CA (1981) Construction of a potential bivalent vaccine strain: introduction of *Shigella sonnei* form I antigen genes into the *galE Salmonella typhi* Ty21a typhoid vaccine strain. Infect Immun 34: 746–750

Formal SB, Hale TL, Kapfer C, Cogan JP, Snoy PJ, Chung R, Wingfield ME, Elisberg BL, Baron LS (1984) Oral vaccination of monkeys with an invasive *Escherichia coli* K-12 hybrid expressing *Shigella flexneri* 2a somatic antigen. Infect Immun 46: 465–469

Gamian A, Romanowska E (1982) The core structure of *Shigella sonnei* lipopolysaccharide and the linkage between O-specific polysaccharide and the core region. Eur J Biochem 129: 105–109

Gemski P Jr, Formal SB (1975) Shigellosis: an invasive infection of the gastrointestinal tract. In: Schlessinger D (ed) American Society for Microbiology, Washington, DC, pp 165–169

Gemski P Jr, Stocker BAD (1967) Transduction by bacteriophage P22 in nonsmooth mutants of *Salmonella typhimurium*. J Bacteriol 93: 1588–1597

Gemski P Jr, Sheahan DG, Washington O, Formal SB (1972) Virulence of *Shigella flexneri* hybrids expressing *Escherichia coli* somatic antigens. Infect Immun 6: 104–111

Gemski P Jr, Koeltzow DE, Formal SB (1975) Phage conversion of *Shigella flexneri* group antigens. Infect Immun 11: 685–691

Godard C, Hannecart-Pokorni E (1977) Étude d'un locus génétique *rfa* impliqué dans la biosynthése du 'core' du lipopolyoside de la paroi *Shigella flexneri* F6S. Ann Inst Pasteur Microbiol 128A: 19–33

Goldman RC, Joiner K, Leive L (1984) Serum-resistant mutants of *Escherichia coli* O111 contain increased lipopolysaccharide, lack an O-antigen-containing capsule, and cover more of their lipid A core with O antigen. J Bacteriol 159: 877–882

Grossman N, Leive L (1984) Complement activation via the alternative pathway by purified *Salmonella* lipopolysaccharide is affected by its structure but not its O-antigen length. J Immunol 132: 376–385

Hale TL (1990) Hybrid vaccines using *Escherichia coli* as an antigen carrier. Res Microbiol 141: 913–919

Hale TL, Formal SB (1989) Oral *Shigella* vaccines. Curr Top Microbiol Immunol 146: 205–211

Hale TL, Formal SB (1990) Live oral vaccines consisting of *Escherichia coli* or *Salmonella typhi* expressing *Shigella* antigens. In: Woodrow GC, Levine MM (eds) New generation vaccines. Marcel Dekker, New York, pp 667–677

Hale TL, Sansoneetti PJ, Schad A, Austin S, Formal SB (1983) Characterization of virulence plasmids and plasmid associated outer membrane proteins in *Shigella flexneri, Shigella sonnei,* and *Escherichia coli.* Infect Immun 40: 340–350

Hale TL, Guerry P, Seid RC, Kapfer C, Wingfield ME, Reaves CB, Baron LS, Formal SB (1984) Expression of lipopolysaccharide O-antigen in *Escherichia coli* K-12 hybrids containing plasmid and chromosomal genes from *Shigella dysenteriae* 1. Infect Immun 46: 470–475

Hannecart-Pokorni E, Godard C, Beumer J (1976) Chimiotypes de mutants R de *Shigella flexneri* et récepteurs de phages. 1. Etude chimique des lipopolyosides. Ann Microbiol 127B: 3–14

Herrero M, Lorenzo VD, Timmis KN (1990) Transposon vectors containing non-antibiotic resistance selection markers for cloning and stable chromosomal insertion of foreign genes in gram-negative bacteria. J Bacteriol 172: 6557–6567

Higgins AR, Floyd TM, Kader MA (1955) Studies in shigellosis III. A controlled evaluation of a monovalent *Shigella* vaccine in a highly endemic environment. Am J Trop Med Hyg 4: 281–284

Inouya M (1979) What is the outer membrane? In: Inouya M (ed) Bacterial outer membranes. Wiley, New York, pp 1–12

Iseki S, Hamano S (1959) Conversion of type antigen IV in *Shigella flexneri* by bacteriophage. Proc Jpn Acad 35: 407–412

Izhar M, Nuchamowitz Y, Mirelman D (1982) Adherence of *Shigella flexneri* to guinea pig intestinal cells is mediated by mucosal adhesin. Infect Immun 35: 1110–1118

Jann K, Jann B (1984) Structure and biosynthesis of O-antigens. In: Rietschel ET (ed) Chemistry of endotoxin. pp 138–186, Handbook of endotoxin, vol. 1. Elsevier, Amsterdam

Jann K, Westphal O (1975) Microbial polysaccharides. In: Sela M (ed) The antigens. Academic, New York, 3: 1–125

Jansson PE, Lindberg AA, Lindberg B, Wollin R (1981) Structural studies on the hexose region of the core in lipopolysaccharides from enterobacteriaceae. Eur J Biochem 115: 571–577

Jiang XM, Neal B, Santiago F, Lee SJ, Romana LK, Reeves PR (1991) Structure and sequence of the *rfb* (O antigen) gene cluster of *Salmonella* serovar *typhimurium* (strain LT2). Mol Miocrobiol 5: 695–713

Joiner KA, Schmetz MA, Goldman RC, Leive L, Frank MM (1984) Mechanism of bacterial resistance to complement-mediated killing: inserted C5b-9 correlates with killing for *Escherichia coli* O111B4 varying in O-antigen capsule and O-polysaccharide coverage of lipid A core oligosaccharide. Infect Immun 45: 113–117

Joiner KA, Grossman N, Schmetz M, Leive L (1986) C3 binds preferentially to long-chain lipopolysaccharide during alternative pathway activation by *Salmonella montevideo*. J Immunol 136: 710–715

Kadis S, Weinbaum G, Ajl SJ (1971) Microbial toxins, vol 5. Academic, New York

Kauffmann F (1961) Die Bakteriologie der *Salmonella* Species. Munksgaard, Copenhagen

Kauffmann F (1966) The bacteriology of enterobacteriaceae. Munksgaard, Copenhagen, pp 76–80

Kenne L, Lindberg B (1983) Bacterial polysaccharides. In: Aspinall GO (ed) The polysaccharides, vol. 2. Academic, New York, pp 287–363 (Molecular biology series)

Ketyl I, Rauss K, Vertenyl A (1974) Oral immunization against dysentery. Acta Microbiol Acad Sci Hung 21: 81–85

Keusch GT, Bennish ML (1988) Shigellosis. In: Evans AS, Brachman P (eds) Bacterial diseases of humans. Plenum, New York

Kiener PA, Marek F, Rodgers G, Lin PF, Warr G, Desiderio J (1988) Induction of tumor necrosis factor, IFN-gamma, and acute lethality in mice by toxic and non-toxic forms of lipid A.J. Immunol 141: 870–874

Kopecko DJ, Washington O, Formal SB (1980) Genetic and physical evidence for plasmid control of *Shigella sonnei* form I cell surface antigen. Infect Immun 29: 207–214

Levine MM (1982) Bacillary dysentery. Mechanisms and treatment. Med Clin North Am 66: 623

Liang-Takasaki CJ, Mäkelä PH, Leive L (1982) Phagocytosis of bacteria by macrophages: changing the carbohydrate of lipopolysaccharide alters interaction with complement and macrophages. J Immunol 128: 1229–1235

Lindberg AA (1973) Bacteriophage receptors. Annu Rev Microbiol 27: 205–237

Lindberg AA, Karnell A, Stocker BAD, Katakura S, Sweiha H, Reinholt FP (1988) Development of an auxotrophic oral live *Shigella flexneri* vaccine. Vaccine 6: 147–150

Lindberg B, Lonngren J, Romanowska E, Ruden U (1972) Location of O-acetyl groups in *Shigella flexneri* types 3c and 4b lipopolysaccharides. Acta Chem Scand 26: 3808–3810

Linde K, Dentchev V, Bondarenko V (1990) Live *Shigella flexneri* 2a and *Shigella sonnei* I vaccine candidate strains with two attenuating markers. I. Construction of vaccine candidate strains with retained invasiveness but reduced intracellular multiplication. Vaccine 8: 25–29

Loppnow H, Brade H, Dürrbaum I, Dinarello CA, Kusumoto S (1989) IL-1 induction-capacity of defined lipopolysaccharide partial structures. J Immunol 142: 3229–3238

Lüderitz O, Westphal O, Staub AM, Nikaido H (1971) Isolation and chemical and immunological characterization of bacterial lipopolysaccharides. In: Weinbaum G, Kadis S, Ajl SJ (eds) Microbial toxins, vol 4. Academic, New York, pp 145-233

Lugtenberg B, van Alphen L (1983) Molecular architecture and functioning of the outer membrane of *Escherichia coli* and other Gram-negative bacteria. Biochem Biophys Acta 737: 51–115

Maagd RAD, Lugtenberg BJJ (1987) Outer membranes of Gram-negative bacteria. Biochem Soc Trans 15 [Suppl] 54–62

Mäkelä P, Stocker BAD (1984) Genetics of lipopolysaccharide. In: Rietschel ET (ed) Chemistry of endotoxin. Elsevier, Amsterdam, pp 59–136 (Handbook of endotoxin, vol 1)

Mäkelä P, Bradley DJ, Brandis H, Frank MM, Hahn H, Henkel W, Jann K, Morse SA, Robbins RB, Rosenstreich DL, Smith H, Timmis K, Tomasz A, Turner MJ, Wiley DS (1980) Evasion of host defences. In: Smith H, Skehel JJ, Turner MJ (eds) The molecular basis of microbial pathogenicity. Verlag Chemie, Weinheim, pp 175–198

Matsui S (1958) Antigenic changes in the *Shigella flexneri* group by bacteriophage. Jpn J Microbiol 2: 153–158

McGrath BC, Osborn MJ (1991) Localization of the terminal steps of O-antigen synthesis in *Salmonella typhimurium*. J Bacteriol 173: 649–654

Mel DM, Terzin AL, Vuksic L (1965a) Studies on vaccination against bacillary dysentery. I. Immunization of mice against experimental *Shigella* infection. Bull WHO 32: 633–636

Mel DM, Terzin AL, Vuksic L (1965b) Studies on vaccination against bacillary dysentery. III. Effective oral immunization against *Shigella flexneri* 2a in a field trial. Bull WHO 32: 647–655

Mel DM, Gangarosa EJ, Radovanovic ML, Arsic BL, Litoinjenko S (1971) Studies on bacillary dysentery. Bull WHO 45: 457–464

Mills SD, Timmis KN (1988) Genetics of O-antigen polysaccharide biosynthesis in *Shigella* and vaccine development. In: Cabello FC, Pruzzo C (eds) Bacteria, complement and the phagocytic cell. Springer, Berlin Heidelberg New York, pp 21–39

Mills SD, Sekizaki T, Gonazalez-Carrero MI, Timmis KN (1988) Analysis and genetic manipulation of *Shigella* virulence determinants for vaccine development. Vaccine 6: 116–122

Mims CA (1982) The pathogenesis of infectious disease, 2nd edn. Academic, London

Nikaido H, Vaara M (1985) Molecular basis of bacterial outer membrane permeability. Microbiol Rev 49: 1–32

Nurminen M, Wahlström E, Kleemola M, Leinonen M, Saikku P, Mäkela, H (1984) Immunologically related ketodeoxyoctonate-containing structures in *Chlamydia trachomatis*, Re mutants of *Salmonella* species, and *Acinetobacter calcoaceticus* var. *anitratus*. Infect Immun 44: 609–613

Okada N, Sasakawa C, Tobe T, Yamada M, Nagai S, Talukder KA, Komatsu K, Kanegasaki S, Yoshikawa M (1991) Virulence-associated chromosomal loci of *Shigella flexneri* identified by random Tn*5* insertion mutagenesis. Mol Microbiol 5: 187–195

Okamura N, Nakaya R (1977) Rough mutants of *Shigella flexneri* 2a that penetrate tissue culture cells but does not evoke keratoconjunctivitis in guinea pigs. Infect Immun 17: 4–8

Okamura N, Nagei T, Nakaya R, Kondo S, Murakami M, Hisatsune K (1983) HeLa cell invasiveness and O antigen of *Shigella flexneri* as separate and prerequisite attributes of virulence to evoke keratoconjunctivitis in guinea pigs. Infect Immun 39: 505–513

Pegues JC, Chen L, Gordon AW, Ding GL, Coleman WG Jr (1990) Cloning, expression, and characterization of the *Escherichia coli* K-12 *rfa*D gene. J Bacteriol 172: 4652–4660

Penn CW (1983) Bacterial envelope and humoral defences. In: Easmon CSF, Jeljaszewicz J, Brown MRW, Lambert PA (eds) Role of the envelope in the survival of bacteria in infection. Academic, London, pp 109–135 (Medical microbiology, vol 3)

Pluschke G, Mayden J, Achtman M, Levine RP (1983) Role of the capsule and the O antigen in resistance of O18: K1 *Escherichia coli* to complement-mediated killing. Infect Immun 42: 907–913

Popoff, MY, Le Minor L (1985) Expression of antigenic factor O54 is associated with the presence of a plasmid in *Salmonella*. Ann Inst Pasteur Microbiol [B] 136: 169–179

Raetz CRH (1990) Biochemistry of endotoxins. Annu Rev Biochem 59: 129–170

Reske K, Jann K (1972) The O8 antigen of *Escherichia coli*. Structure of the polysaccharide chain. Eur J Biochem 67: 320–328

Riley LW, Junio LN, Libaek LB, Schoolnik GK (1987) Plasmid-encoded expression of lipopolysaccharide O-antigenic polysaccharide in enteropathogenic *Escherichia coli*. Infect Immun 55: 2052–2056

Riley LW, Junio LN, Schoolnik GK (1990) HeLa cell invasion by a strain of enteropathogenic *Escherichia coli* that lacks the O-antigenic polysaccharide. Mol Microbiol 4: 1661–1666

Robbins PW, Wright A (1971) Biosynthesis of O-antigens. In: Weinbaum G, Kadis, S, Ajl SJ (eds) Microbial toxins vol 4. Academic, New York, pp 351–368

Robbins PW, Wright A, Dankert M (1966) Polysaccharide biosynthesis. J Gen Physiol 49: 331–346
Romanowska E, Reinhold V (1973) 2-Amino-2-deoxyhexuronic acid: a constituent of *Shigella sonnei* phase I lipopolysaccharide. Eur J Biochem 36: 160–166
Sansonetti PJ, Arondel J (1989) Construction and evaluation of a double mutant of *Shigella flexneri* as a candidate for oral vaccination against shigellosis. Vaccine 7: 443–450
Sansonetti P, David M, Toucas M (1980) Correlation entre la perte d'ADN plasmidique et le passage de la phase I virulente á la phase II avirulente chez *Shigella sonnei.* C R Acad Sci 290: 879–882
Sansonetti P, Formal SB, Hale TL, Kopecko DJ (1981a) Bases genetique de la penetration de *Shigella flexneri* dans les cellules epitheliales. Ann Immunol 132: 183–189
Sansonetti P, Kopecko DJ, Formal SB (1981b) *Shigella sonnei* plasmids: evidence that a large plasmid is necessary for virulence. Infect Immun 34: 75–83
Sansonetti PJ, Hale TL, Dammin GJ, Kapfer C, Collins HH Jr, Formal SB (1983) Alterations in the pathogenesis of *Escherichia coli* K-12 after transfer of plasmid and chromosomal genes from *Shigella flexneri.* Infect Immun 34: 1392–1402
Seid RC, Kopecko DJ, Sadoff JC, Schneider H, Baron LS, Formal SB (1984) Unusual lipopolysaccharide antigens of *Salmonella typhi* and vaccine strain expressing the *Shigella sonnei* form I antigen. J Biol Chem 259: 9028–9034
Simmons DAR (1971) Immunochemistry of *Shigella flexneri* O-antigens: a study of structural and genetic aspects of the biosynthesis of cell-surface antigens. Bacteriol Rev 35: 117–148
Sturm S, Timmis KN (1986) Cloning of the *rfb* gene region of *Shigella dysenteriae* 1 and construction of an *rfb-rfp* gene cassette for the development of lipopolysaccharide-based live anti-dysentery vaccines. Microb Pathog 1: 289–297
Sturm S, Fortnagel P, Timmis KN (1984) Immunoblotting procedure for the analysis of electrophoretically fractionated bacterial lipopolysaccharides. Arch Microbiol 140: 198–201
Sturm S, Jann B, Jann K, Fortnagel P, Timmis KN (1986a) Genetic and biochemical analysis of *Shigella dysenteriae* 1 O antigen polysaccharide biosynthesis in *Escherichia coli* K-12: 9 kb plasmid of *S. dysenteriae* 1 determines addition of a galactose residue to the lipopolysaccharide core. Microb Pathog 1: 299–306
Sturm S, Jann B, Jann K, Fortnagel P, Timmis KN (1986b) Genetic and biochemical analysis of *Shigella dysenteriae* 1 O antigen polysaccharide biosynthesis in *Escherichia coli* K-12: structure and functions of the *rfb* gene cluster. Microb Pathog 1: 307–324
Subbaiah TV, Stocker BAD (1964) Rough mutants of *Salmonella typhimurium.* Nature (London) 201: 1298–1299
Tacket CO, Forrest B, Morona R, Attridge SR, Labroody J, Tall BD, Reymann M, Rowley D, Levine MM (1990) Safety, immunogenicity, and efficacy against cholera challenge in humans of a typhoid-cholera hybrid vaccine derived from *Salmonella typhi* Ty21a. Infect Immun 58: 1620–1627
Timakov VD, Petrovskava VG, Bondarenko VM (1970) Studies of the genetic control of *Shigella* subgroup B type specific antigens. Ann Inst Pasteur 118: 3–9
Timmis KN, Boulonois GJ, Bitter-Suermann D, Cabello FC (1985) Surface components of *Escherichia coli* that mediate resistance to bactericidal activities of serum and phagocytosis. Curr Top Microbiol Immunol 118: 197–218
Timmis KN, Gonzalez-Carrero MI, Sekizaki T, Rojo F (1986a) Biological activities specified by antibiotic resistance plasmids. J Antimicrob Chemother 18 (Suppl. C): 1–12
Timmis KN, Sturm S, Watanabe H (1986b) Genetic dissection of pathogenesis determinants of *Shigella* and enteroinvasive *Escherichia coli.* In: Holmgren J, Lindberg AA, Möllby R (eds) Development of vaccines and drugs against diarrhoea. Student Literatur, Lund, pp 107–126
Verma NK, Reeves PR (1989) Identification and sequence of *rfb*S and *rfb*E, which determine antigenic specificity of group A and group D salmonellae. J Bacteriol 171: 5694–5701
Verma NK, Quigley NB, Reeves PR (1988) O-antigen variation in *Salmonella* spp.: *rfb* gene clusters of three strains. J Bacteriol 170: 103–107
Watanabe H, Timmis KN (1984) A small plasmid in *Shigella dysenteriae* 1 specifies one or more functions essential for O antigen production and bacterial virulence. Infect Immun 43: 391–396
Watanabe H, Nakamura A, Timmis KN (1984) Small virulence plasmid of *Shigella dysenteriae* 1 strain W30864 encodes a 41,000-dalton protein involved in formation of specific lipopolysaccharide side chains of serotype 1 isolates. Infect Immun 46: 55–63
Westphal O, Jann K, Himmelspach K (1983) Chemistry and immunochemistry of bacterial lipopolysaccharides as cell wall antigens and endotoxins. Prog Allergy 33: 9–39

Wollenweber H-W, Rietschel ET (1990) Analysis of lipopolysaccharide (lipid A) fatty acids. J Microbiol Methods 11: 195–211

Wolpe SD, Davatelis G, Sherry B, Beutler B, Hesse DG (1988) Macrophages secrete a novel heparin-binding protein with inflammatory and neutrophil chemokinetic properties. J Exp Med 167: 570–581

Wyk P, Reeves PR (1989) Identification and sequence of the gene for abequose synthetase, which confers antigenic specificity on group B salmonellae: homology with galactose epimerase. J Bacteriol 171: 5687–5693

Yoshida Y, Okamura N, Kato J, Watanabe H (1991) Molecular cloning and characterization of form I antigen genes of *Shigella sonnei*. J Gen Microbiol 137: 867–874

Shiga Toxin: Biochemistry, Genetics, Mode of Action, and Role in Pathogenesis

A. D. O'BRIEN[1], V. L. TESH[1], A. DONOHUE-ROLFE[2], M. P. JACKSON[3], S. OLSNES[4], K. SANDVIG[4], A. A. LINDBERG[5], and G. T. KEUSCH[2]

1 Introduction

Dysentery was well known and clearly described in many ancient texts and histories. The first step towards the description of the genus *Shigella*, however, was the identification of *Entamoeba histolytica* by LOSCH in 1875 and the

[1] Department of Microbiology, F. Edward Hébert School of Medicine, Uniformed Services University of the Health Sciences, 4301 Jones Bridge Road, Bethesda, MD 20814–4799, USA
[2] Department of Medicine, Division of Geographic Medicine and Infectious Diseases, Box 41, New England Medical Center Hospitals, Tufts University School of Medicine, 750 Washington Street, Boston, MA 02111, USA
[3] Department of Immunology and Microbiology, Wayne State University School of Medicine, 540 East Canfield Avenue, Detroit, MI 48201, USA
[4] Department of Biochemistry, Institute for Cancer Research, Montebello, 0310 Oslo 3, Norway
[5] Department of Clinical Bacteriology, Huddinge University Hospital, 141 86 Huddinge, Sweden

Current Topics in Microbiology and Immunology, Vol. 180

separation of amebic from all other forms of dysentery (LOSCH, 1875). With this discovery, attention could be focused on the etiology of epidemic dysentery, and a partial description of the prototype *Shigella* sp., *Shigella dysenteriae* type 1, was published by CHANTEMESSE and WIDAL in 1888. The definitive description of this organism was provided by Kiyoshi Shiga following an extensive dysentery epidemic in Japan in 1896 (SHIGA 1898). It did not take long to determine that there was a potent toxic activity in this organism, and in 1900 FLEXNER reported that either living or killed cultures of Shiga's bacillus injected into the peritoneal cavity of animals caused fever and diarrhea. Flexner concluded that shigellosis was due to a "toxic agent rather than to an infection per se"; however, the observed effects were probably due to endotoxin. The presence of a lethal toxin in extracts of heat-killed bacteria was shown independently by NEISSER and SHIGA (1903) and by CONRADI (1903). CONRADI (1903) also described the limb paralysis following parenteral inoculation of *Shigella* extracts in rabbits, characteristic of the so-called Shiga neurotoxin.

During the next 50 years research focused on the separation of endotoxin, clearly responsible for some of the effects described above, from the protein toxin causing the neurological symptoms (KEUSCH et al. 1986a). By 1953, when VAN HEYNINGEN and GLADSTONE (1953a) achieved significant purification of the protein toxin, they concluded that toxin played no role in the pathogenesis of shigellosis. In the next decade, two important findings were reported: first, that Shiga neurotoxin appeared to target vascular endothelium (BRIDGWATER et al. 1955; HOWARD 1955); and secondly, that toxin was lethal to certain cells in culture (VICARI et al. 1960).

The modern era in the study of Shiga toxin began in 1972, when it was reported that inoculation of Shiga toxin in ligated rabbit ileal loops caused inflammatory enteritis and resulted in net fluid accumulation (KEUSCH et al. 1972a). In the course of screening *Escherichia coli* isolates for heat-labile enterotoxin production using Vero cells, an African green monkey kidney cell line, KONOWALCHUK et al. (1977, 1978) noted that some strains of *E. coli*, isolated from human and animal sources, possessed an irreversible cytotoxic activity. Antibody neutralization and molecular biology studies showed that the cytotoxins were antigenically and genetically related to Shiga toxin (O'BRIEN et al. 1982; STROCKBINE et al. 1986, 1988), leading to the designation "Shiga-like" toxins (SLTs).

In the past 10 years, the genes for these toxins have been identified and sequenced, the toxins have been purified to homogeneity, their biochemical mechanism of action as well as the tissue receptor to which they bind have been discovered, and their physiological role in pathogenesis has been explored. This chapter summarizes the enormous progress in our understanding of this family of toxins.

2 Structure of Shiga Toxin and the SLTs

Members of the Shiga toxin family are bipartite molecules composed of a single enzymatic A subunit and a multimer of receptor-binding B subunits (DONOHUE-ROLFE et al. 1984, 1989b; OLSNES et al. 1981). Amino acid sequence comparison between Shiga toxin, produced by *S. dysenteriae* type 1, and the SLTs (SLT-I, SLT-II, and SLT-IIv), produced by *E. coli*, revealed that the sizes of the mature A and B polypeptides, as well as the position of a single intrachain disulfide bond in each subunit, are highly conserved (CALDERWOOD et al. 1987; DEGRANDIS et al. 1987; JACKSON et al. 1987a, b; SEIDAH et al. 1986; STROCKBINE et al. 1988; WEINSTEIN et al. 1988b). Furthermore, the observation that the A and B subunits of Shiga toxin/SLT-I, SLT-II, and SLT-IIv can reassemble into functional hybrid cytotoxins (ITO et al. 1988; WEINSTEIN et al. 1989) indicates that the tertiary conformation of the individual polypeptides is also highly conserved. Although the heat-labile enterotoxins of *Vibrio cholerae* and *E. coli* (GILL 1976) are composed of a single enzymatic subunit and five binding subunits, the unique enzymatic activity and receptor specificity of the Shiga toxin family distinguishes them from other bacterial cytotoxins (DEGRANDIS et al. 1987; ENDO et al. 1988; JACEWICZ et al. 1986; LINDBERG et al. 1987; LINGWOOD et al. 1987; SAMUEL et al. 1990; WADDELL et al. 1988).

As with other bacterial cytotoxins, the A polypeptides of the Shiga toxin family are activated by proteolytic processing. A bacterial protease nicks the Shiga toxin and SLT A subunits into an enzymatic A1 fragment (approximately 27 kDa) and a carboxyl terminal A2 fragment (approximately 4 kDa) which remain linked by a single disulfide bond until the enzymatic fragment is released and enters the cytosol of a susceptible mammalian cell (OLSNES et al. 1981). Amino terminal sequence analysis of the purified Shiga toxin A subunit located the nicking site between alanine 253 and serine 254 of the 293 amino acid mature polypeptide (TAKAO et al. 1988). Biochemical studies suggest that the B polypeptides are noncovalently associated with the A2 portion of the A subunit in the holotoxin, although free B subunits are capable of forming pentamers which can block receptor binding (CALDERWOOD et al. 1990; DONOHUE-ROLFE et al. 1989b). Keusch and coworkers (KEUSCH 1981; MOBASSALEH et al. 1988) proposed the existence of an entry domain responsible for mediating translocation of the A1 fragment out of the endocytotic vesicle into the cytoplasm. Although this function has not been localized to either polypeptide, hydropathy profiles of the A2 fragment (JACKSON et al. 1987a) and the B subunit (SEIDAH et al. 1986) reveal hydrophobic regions which may be important for translocation.

Purified B subunit of Shiga toxin/SLT-I has been recently crystallized in a form suitable for high-resolution X-ray analysis (HART et al. 1991). Although a pentameric arrangement of B subunits is supported by biochemical cross-linking data, preliminary X-ray data of the crystals display fourfold and not fivefold symmetry consistent with a tetrameric arrangement of the B subunits. Definitive subunit stoichiometry will have to await crystallization of the holotoxin.

3 Toxin Purification

The amount of Shiga toxin produced by strains of *S. dysenteriae* type 1 is heavily dependent upon the bacterial culture conditions. Under optimum culture conditions, Shiga toxin comprises approximately 0.1% of the total protein, but the initial yields can be many times lower if less than optimum conditions are used. There are several known culture conditions which improve the initial yield of toxin. First, since toxin production is controlled by the level of iron (DUBOS and GEIGER 1946; VAN HEYNINGEN and GLADSTONE 1953b; WEINSTEIN et al. 1988a), the *S. dysenteriae* type 1 strain should be grown in low-iron medium such as modified syncase broth (KEUSCH et al. 1988; O'BRIEN and LAVECK 1983). The optimum iron level is approximately 0.1 μg Fe^{3+}/ml (KEUSCH et al. 1986a; VAN HEYNINGEN and GLADSTONE 1953b); below this level, although toxin-specific activity may increase, the bacterial growth is partially inhibited. Secondly, growth at 37 °C is optimal for toxin production (WEINSTEIN et al. 1988a). Growth at lower temperatures has been shown to reduce the overall yield of toxin. Finally, toxin production is apparently reduced under anaerobic conditions (DUBOS and GEIGER,1946). Therefore, in general, cultures are grown aerobically using standard laboratory shakers.

The strain of *S. dysenteriae* used for toxin production does not seem to be of critical importance. With the exception of a few constructed Tox^- strains, all strains of *S. dysenteriae* type 1 seem to produce Shiga toxin, and among these strains there are no reported quantitative differences in the level of toxin produced. Because of the impressive virulence of the organism, most laboratories, owing to safety concerns, have used avirulent strains such as the rough, noninvasive mutant strain 60R first described by DUBOS and GEIGER (1946).

In 1980, the first successful purification scheme for Shiga toxin was published (OLSNES and EIKLID 1980). Since then, Shiga toxin has been purified by a variety of schemes utilizing a variety of chromatographic methods (BROWN et al. 1982; DONOHUE-ROLFE et al. 1984, 1989b; O'BRIEN et al. 1980; O'BRIEN and LAVECK 1983; OLSNES et al. 1981; RYD et al. 1989; YUTSUDO et al. 1986). These methods include molecular sieve, ion exchange chromatography, chromato- and isoelectric focusing, and various forms of affinity chromatography. The various schemes naturally vary dramatically with respect both to their final yields and total recoveries of toxin, and to their ease of execution. This review will outline briefly one straightforward purification scheme which uses three conventional chromatography steps with total toxin recoveries of about 50%. Full details of this purification procedure have been published elsewhere (DONOHUE-ROLFE et al. 1984; KEUSCH et al. 1988).

Cultures of *S. dysenteriae* type 1 strain are grown aerobically in a low-iron medium at 37 °C. Since toxin is produced during the logarithmic phase of growth, bacteria are harvested when stationary growth is reached. Shiga toxin is a cell-associated protein located in the periplasmic space (DONOHUE-ROLFE and KEUSCH 1983), and hence maximum toxin yields are obtained by making a lysate of the bacteria. The crude cell lysate is first applied in low ionic strength buffer to a column containing the dye Cibacron Blue F3G-A coupled to Sepharose (Blue Sepharose). Shiga toxin and approximately 10% of the other proteins bind to the Blue Sepharose and after extensive washing are eluted using higher ionic strength buffer. The partially purified toxin is then subjected to chromatofocusing. Shiga toxin elutes from the chromatofocusing column as a well-defined peak, as determined by A_{280}, over an elution buffer pH range of 7.0–7.1. The final step in the purification scheme is molecular sieve chromatography. By gel filtration, Shiga toxin elutes with an apparent molecular weight of 45000. Compared to the crude bacterial lysate, the toxin purified by this three-step procedure (Fig. 1) results in about a 1300-fold increase in toxin-specific activity, and about 1 mg toxin is obtained from 3 L culture.

Recently alternative toxin purification schemes have been developed which take advantage of the ability of Shiga toxin to bind tightly to carbohydrates containing terminal galα1-4gal (DONOHUE-ROLFE et al. 1989b; RYD et al. 1989). One of these schemes utilizes glycoproteins from hydatid cysts that possess P_1 blood group reactivity (DONOHUE-ROLFE et al. 1989b). The glycoproteins contain the terminal trisaccharide galα1-4galβ1-4glc, which is a toxin receptor.

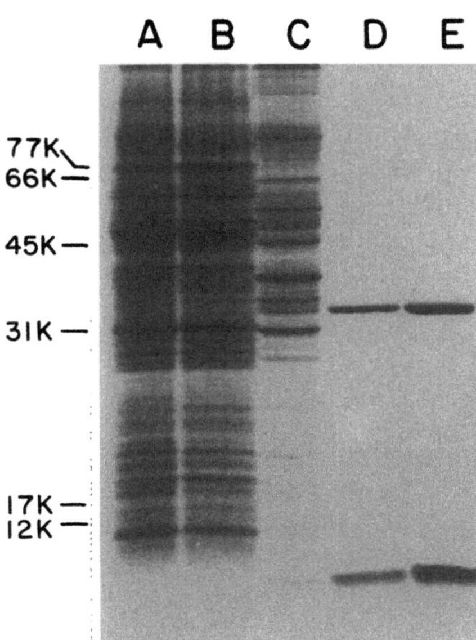

Fig. 1. Sodium dodecyl sulfate (SDS) polyacrylamide gel electrophoresis of Shiga toxin during purification by the method described in the text. Samples were dissolved in SDS sample buffer containing 2-mercaptoethanol, heated in boiling water for 10 min, and applied to a 15% polyacrylamide gel. *Lane A*, crude lysate of *S. dysenteriae* type 1; *lane B*, Blue Sepharose flow-through; *lane C*, Blue Sepharose salt eluate; *lane D*, pH 7.1 fraction from chromatofocusing step; *lane E*, molecular sieve-purified toxin

By coupling partially purified preparations of the glycoproteins to Sepharose 4B, a toxin affinity matrix is generated. Shiga toxin binds very strongly to the matrix which allows the use of a high ionic strength buffer to completely remove weakly associated contaminating proteins. Shiga toxin can be removed from this receptor analogue affinity column by eluting with 4.5 *M* $MgCl_2$. The toxin purification procedure described by RYD et al. (1989) involves the covalent coupling of the trisaccharide galα1-4galβ1-4glc to polyvinyl/polyacyl gel (Fractogel). Shiga toxin is eluted from this receptor analogue matrix with 6 *M* guanidine HCl, followed by immediate dialysis. These affinity chromatography purification protocols allow one-step purification of Shiga toxin with total toxin recoveries of greater than 80%. They also have the advantage that, since they are based on the recognition of the carbohydrate receptor, they can be applied to the purification of the B subunit and to any of the SLTs that display similar binding specificities (CALDERWOOD et al. 1990; DONOHUE-ROLFE et al. 1989b).

4 Genetics of Shiga Toxin and the SLTs

4.1 Genetic Loci of the Shiga Toxin and SLT Genes

4.1.1 Chromosomal Locus of the Shiga Toxin Operon

Originally, the genetic determinant which encodes Shiga toxin was mapped near a locus on the *S. dysenteriae* type 1 chromosome associated with the provocation of fluid accumulation in ligated ileal loops (TIMMIS et al. 1985). Subsequently, the Shiga toxin operon (designated *stx*) was correctly mapped near *pyrF*, a chromosomal locus at 28 min which is not associated with other virulence determinants of *Shigella spp.* (SEKIZAKI et al. 1987). Isolation of the *stx* operon facilitated the development of nontoxigenic strains of *S. dysenteriae* used to investigate pathogenesis. In 1972, GEMSKI et al. described spontaneous chlorate-resistant mutants of *S. dysenteriae* type 1 which did not produce Shiga toxin. Although the genetic mechanism for the loss of toxin production was not known, these chlorate-resistant strains were used in animal and volunteer studies to assess the role of Shiga toxin in disease (GEMSKI et al. 1972; LEVINE et al. 1973). Subsequently, NEILL et al. (1988) characterized these mutants by demonstrating that the coincident acquisition of chlorate resistance and the loss of Shiga toxin production were caused by a chromosomal deletion encompassing the *chl* locus (at 27 min) and the *stx* operon. In addition to the chlorate-resistant mutants, nontoxigenic derivatives of *S. dysenteriae* type 1 constructed by transposon mutagenesis of the *stx* operon have been used in animal studies (FONTAINE et al. 1988; NEILL et al. 1988; SEKIZAKI et al. 1987).

4.1.2 SLT-I and SLT-II are Bacteriophage Encoded

Strains of *E. coli* which cause edema disease (ED) in swine and enterohemorrhagic *E. coli* (EHEC) which cause hemorrhagic colitis in humans produce a family of cytotoxins with biological activities similar to Shiga toxin. Prototypes of the SLT family are: SLT-I which is neutralized by antiserum to Shiga toxin (O'BRIEN and HOLMES 1987), the antigenically distinct SLT-II (STROCKBINE et al. 1986), and a variant of SLT-II (SLT-IIv) (LINGGOOD and THOMPSON 1987; MARQUES et al. 1987). The transmissible nature of SLT production among *E. coli* was first reported by SMITH and LINGGOOD (1971), although they were unable to demonstrate plasmid involvement. Later, several investigators established the role that lysogenic conversion has in SLT production using bacteriophage preparations induced from various strains of EHEC (O'BRIEN et al. 1984; SCOTLAND et al. 1983; SMITH et al. 1983; STROCKBINE et al. 1986). While SLT-I and SLT-II production may be encoded by bacteriophage, SLT-IIv appears to be encoded from a chromosomal locus in ED strains with no evidence of lysogenic conversion (MARQUES et al. 1987; RIETRA et al. 1989; SMITH et al. 1983).

The first SLT-converting prophages to be characterized were isolated from the EHEC strains H19 (O26:H11) and 933 (O157:H7). Strain H19 is lysogenized by two SLT-I-converting bacteriophages, designated H19A and H19B, which possess different host ranges (SMITH et al. 1983). H19B is related to coliphage lambda and carries the *slt-I* operon far from the attachment site, indicating that it became bacteriophage-associated in the distant past (HUANG et al. 1987). Strain 933 was originally characterized as a double lysogen with two morphologically distinct bacteriophages designated 933J, which encodes SLT-I, and 933W, which encodes SLT-II (O'BRIEN et al. 1984; STROCKBINE et al. 1986) and has the capacity to transduce SLT-I at a low frequency (O'BRIEN et al. 1989). However, subsequent reports demonstrated that 933J and H19A are independent isolates of the same bacteriophage and although hybridization analysis shows that 933 carries H19A-related sequences, only bacteriophage 933W can be induced from this strain (O'BRIEN et al. 1989; WILLSHAW et al. 1987). Perhaps the *slt-I* operon is associated with a defective H19A-related bacteriophage in strain 933 which can be rescued by 933W (O'BRIEN et al. 1989).

In summary, the O26 and O157 EHEC serogroups carry at least two morphologically distinct families of SLT-encoding prophages which are represented by H19A and 933W (RIETRA et al. 1989). In contrast, neither Shiga toxin production by *S. dysenteriae* type 1 nor SLT-IIv production by ED strains is subject to lysogenic conversion. DNA hybridization analysis using bacteriophage-specific probes revealed the presence of H19A- and 933W-related sequences in several EHEC and ED strains but not in *Shigella* spp. (NEWLAND and NEILL 1988). Perhaps bacteriophages are responsible for the dissemination of Shiga toxin/SLT genes within the *E. coli* gene pool but not among the shigellae.

4.2 Cloning and Sequence Analysis

4.2.1 Cloning of the Shiga Toxin and SLT Structural Genes

Prior to the discovery that SLT antigenic variants exist, the *slt-I* operon was isolated from the O26:H11 strain H19 by WILLSHAW et al. (1985) and from the O157:H7 strain 933 by NEWLAND et al. (1985). In that same report by WILLSHAW et al., the structural genes for an antigenically distinct SLT were isolated from a prophage of the O157:H EHEC strain E32511. Subsequent studies (SCHMITT et al. 1991; STROCKBINE et al. 1986) demonstrated that strain E32511 produces an SLT-II which is nearly identical to the bacteriophage 933W-encoded toxin, as well as a variant of SLT-II. DNA hybridization analysis revealed that the *slt-I* operon isolated from strains H19 and 933 and the *slt-II* operon isolated from strains E32511 and 933 share limited sequence homology (NEWLAND et al. 1985, 1987; WILLSHAW et al. 1985). This observation was later supported by nucleotide sequence analysis (see below).

While all *slt-I* isolates examined thus far are essentially identical, there is considerable heterogeneity in the *slt-II* gene family. The structural genes for SLT-IIv were first isolated from the chromosome of an *E. coli* strain which causes ED in pigs (GYLES et al. 1988; WEINSTEIN et al. 1988b). Subsequently, the genes for additional variants of SLT-II have been isolated from EHEC (GANNON et al. 1990; ITO et al. 1990; SCHMITT et al. 1991). A sequence comparison of the growing *slt-II* family suggests that genetic recombination among the B subunit genes, rather than base substitutions, has given rise to the variants of SLT-II found in human and animal strains of *E. coli* (GANNON et al. 1990; ITO et al. 1990). Perhaps natural selection has diversified the *slt-II* gene family, while the *slt-I* genes have remained relatively unchanged. On the other hand, because SLT-II-producing *E. coli* are more frequently isolated as pathogens of humans and livestock, the distribution of clinical isolates available for genetic analysis may be biased.

Although the *stx* operon had been isolated on a conjugative plasmid for mapping studies (SEKIZAKI et al. 1987), there were restrictions prohibiting the use of a cloning system which might yield a recombinant producing high levels of Shiga toxin. When these restrictions were lifted, the *stx* operon was isolated from *S. dysenteriae* 1 on a high copy number vector in *E.coli,* and, as discussed below, sequence analysis demonstrated that Shiga toxin and SLT-I are essentially identical (STROCKBINE et al. 1988). Although KOZLOV et al. (1988) demonstrated that the *stx* operon is flanked by insertion sequences in strain 60R, there have been no reports showing multiple copies of the *stx* operon in the chromosome of *S. dysenteriae* type 1. Insertion sequences may be responsible for duplication of the *slt-II* genes in EHEC strains such as E32511, although there is no evidence to support this and the role of lysogenic conversion must also be considered. Finally, despite the efforts of many investigators, the genetic determinant for low-level SLT production has never been isolated from *Shigella* spp., *E. coli,* or *V. cholerae.*

4.2.2 Sequence Analysis of Shiga Toxin and SLTs

The nucleotide and amino acid sequences have been reported for Shiga toxin (KOZLOV et al. 1988; SEIDAH et al. 1986; STROCKBINE et al. 1988), SLT-I (CALDERWOOD et al. 1987; DEGRANDIS et al. 1987; JACKSON et al. 1987b), SLT-II (JACKSON et al.1987a), and SLT-IIv (GYLES et al. 1988; WEINSTEIN et al. 1988b). A sequence comparison (STROCKBINE et al. 1988) revealed that the *stx* and *slt-I* operons are essentially identical and that Shiga toxin and SLT-I differ by a single conservative amino acid change in the A subunits. The *slt-II* and *slt-IIv* operons are approximately 90% homologous to each other and 55% homologous to the *stx*/*slt-I* operon (JACKSON et al. 1987a; WEINSTEIN et al. 1988b). The operons for every member of the Shiga toxin family are organized identically: the A and B subunit genes are arranged in tandem with the A subunit gene promoter-proximal and separated from the B subunit gene by a 12–15 nucleotide gap. A promoter has been identified 5′ to the A subunit genes and the existence of an independent promoter for the B subunit genes has been suggested. Finally, sequence analysis identified putative ribosome binding sites which precede both the A and B subunit genes.

The mature A subunit of Shiga toxin is composed of 293 amino acids with a calculated molecular weight of 32225 (STROCKBINE et al. 1988), and the mature B subunit is composed of 69 amino acids with a molecular weight of 7691 (SEIDAH et al. 1986). Sequence analysis has illustrated the amount of conformational similarity which exists among members of the Shiga toxin family. The sizes of the mature A and B polypeptides of Shiga toxin/SLT-I, SLT-II, and SLT-IIv are very similar, and the number and placement of cysteine residues are identical (JACKSON et al. 1987a; WEINSTEIN et al. 1988b). There are two regions of limited sequence homology between the A subunits of all members of the Shiga toxin family and the plant toxin ricin (CALDERWOOD et al. 1987; DEGRANDIS et al. 1987; JACKSON et al. 1987a; STROCKBINE et al. 1988; WEINSTEIN et al. 1988b) which suggests that the structure of these cytotoxins has been well conserved for a common function. This hypothesis is supported by the observation that Shiga toxin, the SLTs, and ricin inhibit eukaryotic protein synthesis by an identical mechanism of action. As discussed below, the discovery that these cytotoxins from diverse sources share a common mode of action and limited sequence homology has substantially facilitated structure-function studies.

4.3 Genetic Regulation

Production of Shiga toxin by *S. dysenteriae* type 1 is repressed by high levels of iron and reduced temperature (O'BRIEN and HOLMES 1987; WEINSTEIN et al. 1988a). Interestingly, other attributes of *Shigella* virulence, such as the ability to invade epithelial cells, are also temperature regulated (MAURELLI et al. 1984). SLT-I production by *E. coli* is also reduced by high levels of iron, but is unaffected by temperature, while SLT-II and SLT-IIv production is unaffected by either iron or temperature (SUNG et al. 1990). These findings suggest that expression of the virulence determinants of *S.dysenteriae* and EHEC is coordinately regulated in response to the environmental signals encountered in the human host, i.e., 37 °C and reduced levels of iron.

Primer extension analysis has been used to map promoters for the *stx/slt-I* (DEGRANDIS et al. 1987), *slt-II*, and *slt-IIv* (SUNG et al. 1990) operons 5′ to the A subunit genes. Nucleotide sequence analysis revealed that the promoter of the *stx/slt-I* operon contains a region of dyad symmetry which is found in the promoters of other iron-repressed genes (BETLEY et al. 1986; DEGRANDIS et al. 1987). CALDERWOOD and MEKALANOS (1987, 1988) definitively established that the *stx/slt-I* operon is regulated by the *fur* gene product, a DNA-binding protein which complexes with iron and blocks transcription. Because the *slt-II* and *slt-IIv* promoters lack the Fur operator sequence, the production of SLT-II and SLT-IIv are unaffected by the concentration of iron in the media (SUNG et al. 1990).

The 5B:1A subunit composition of Shiga toxin and the SLTs (DONOHUE-ROLFE et al. 1984) suggests that A and B polypeptide production may be independently regulated. Several investigators (DEGRANDIS et al. 1987; HUANG et al. 1986; KOZLOV et al. 1988; NEWLAND et al. 1985) have provided evidence that the *slt-IB* gene may be expressed from an independent promoter which lies in the 3′ sequences of the *slt-IA* gene. Although the existence of an *slt-IIB* gene promoter could not be established (SUNG et al. 1990), the findings do not rule out the existence of an independent promoter for the *stxB* gene, but do strongly support the role of an independent ribosome binding site in Shiga toxin B subunit synthesis.

In summary, repression of the *stx/slt-I* operon by iron is governed by Fur, and expression of the B subunit gene may be regulated by an independent promoter, ribosome binding site, or both. The roles of regulation by iron and the independent expression of the subunit genes on pathogenesis are unknown.

5 Receptors

The study of toxin-cell interactions using cultured cells has led to the identification of a toxin-binding membrane glycolipid, globotriaosylceramide (or Gb_3), which appears to be the principle functional receptor for Shiga toxin. Initial attempts to define a specific cell surface toxin receptor predate the preparation of purified radiolabeled Shiga toxin (KEUSCH and JACEWICZ 1977). These experiments revealed that only sensitive, and not resistant, cells were able to remove toxin bioactivity from the overlying medium, suggesting that binding and removal of the toxin molecule from the fluid phase was occurring. Other data implicated cell surface glycoproteins containing oligomeric β1-4-linked *N*-acetyl-*D*-glucosamine in this process. When tunicamycin, an inhibitor of *N*-linked glycoprotein synthesis, was observed to inhibit toxin activity and reduce toxin binding to HeLa cells, it was reasonably concluded that toxin bound to a glycoprotein receptor via a sugar-specific mechanism (KEUSCH et al. 1986b), which is commonly mediated by glycoprotein receptors.

With the subsequent availability of highly purified iodinated toxin, it became possible to carry out direct binding studies, both in tissue culture and with microvillus membranes (MVM) from toxin-responsive rabbit small intestine. Using classical receptor analysis methods, FUCHS et al. (1986) demonstrated that toxin bound to MVM in a rapid, reversible, specific, and temperature-dependent manner. Scatchard analysis showed a single class of binding site, with an equilibrium association constant, K, of $4.7 \times 10^9 M^{-1}$. At 4 °C, the calculated number of toxin molecules binding to the MVM was 7.9×10^{10}, equivalent to 1.2×10^6 toxin molecules bound per enterocyte. When similar methods were applied to cloned sensitive and resistant HeLa cells, however, two classes of binding sites were found using a computer-based Scatchard model program (JACEWICZ et al. 1989). The two receptors were a low-affinity, high-capacity site, and a high-affinity, low-capacity site. The number of high-affinity sites (but not of low-affinity sites) correlated with the sensitivity of the cell line to the action of the toxin. Treatment of HeLa cells with tunicamycin led to a time-dependent decrease in the number of high-affinity sites, consistent with the postulated role of the high-affinity binding site as a functional receptor.

However, in 1983, BROWN et al. reported that Shiga toxin bound to glycosphingolipids possessing terminal galα1-4gal disaccharides, such as Gb_3, consisting of a trisaccharide (galα1-4galβ1-4glu) β-linked to ceramide. JACEWICZ et al. (1986) continued to look for the hypothesized receptor glycoprotein, but identified only a glycolipid toxin-binding fraction in rabbit MVM. When this was extracted from either MVM or toxin-sensitive HeLa cells, it was determined to be Gb_3 by comparing its migration to authenticated glycolipid standards on thin-layer chromatograms. Two other digalactosyl toxin-binding glycosphingolipids were also detected, galabiosylceramide in HeLa cells and P1 blood group antigen from human type B erythrocyte membranes. The identity of the HeLa- and MVM-binding sites was subsequently confirmed by high-performance liquid chromatography (HPLC) of benzoylated neutral glycolipids (MOBASSALEH et al. 1989), and the MVM receptor was found to contain only hydroxylated fatty acids in the lipid moiety.

LINDBERG et al. (1987) followed up the preliminary report of BROWN et al. (1983) by an extensive comparison of toxin binding to isolated purified glycolipids and to HeLa cells. The presence of the terminal galα1-4gal disaccharide was essential for binding of toxin, whereas glycolipids with internal galabiose were inactive. Unless covalently coupled to bovine serum albumin (BSA), however, soluble free galabiose did not competitively inhibit toxin binding to HeLa cells. In fact, the inhibitory effect of galabiose-BSA was directly proportional to the amount of disaccharide coupled to the protein, indicating that binding required multivalency of the carbohydrate moiety. LINDBERG et al. (1987) also suggested that two sites might be involved in toxin binding since some cells densely covered with toxin receptors were resistant to toxin action.

It was suggested that galabiosyl present in glycoproteins might be able to bind but not lead to further processing of the toxin.

It is well recognized that the presence of a binding site does not mean that it necessarily functions as a receptor unless binding relates to some function of the ligand being bound. Several pieces of data suggest that Gb_3 is, in fact, the functional toxin receptor resulting in the fluid secretory (enterotoxin) effects of Shiga toxin in the rabbit small bowel. For example, MOBASSALEH et al. (1988) found that infant rabbits younger than 15 days of age did not respond to lumenal inoculation of Shiga toxin by net secretion of fluids. Only older animals responded in this way, with the amount of fluid accumulating in ligated loops progressively increasing in animals from 16 to 35 days of age. This age-related effect correlated with the content of Gb_3 detected in rabbit intestinal MVM. Thus, while MOBASSALEH et al. (1989) found very low levels of Gb_3 in the MVM from newborn animals, there was a rapid increase in Gb_3 content in tissue from 16- to 35-day-old animals, with a concomitant drop in the content of the Gb_3 precursor lactosylceramide. The same investigators have reported preliminary evidence that developmental regulation of Gb_3 content in the rabbit is due, at least in part, to age-related increases in the microsomal biosynthetic enzyme uridine diphosphate (UDP)-galactose:lactosylceramide galactosyltransferase (MOBASSALEH et al. 1991).

The rabbit intestinal secretion model has provided additional evidence that Gb_3 is the receptor mediating the enterotoxic effects of Shiga toxin. KANDEL et al. (1989) reported that, although Shiga toxin significantly decreases neutral sodium absorption in rabbit jejunum, it does not alter substrate-coupled absorption or active chloride secretion. Since absorption is mediated by the villus cell and secretion by the crypt cell, the physiological data suggest that toxin acts only on the absorptive villus cell. Villus and crypt cells were therefore isolated from rabbit jejunum by a standard method; villus cells alone expressed detectable Gb_3, were able to bind toxin, and responded by an inhibition of leucine incorporation into protein. Thus, Shiga toxin appears to target the villus cell, presumably because it, and not the crypt cell, expresses Gb_3 receptor.

A variety of additional studies in cell culture indicate that Gb_3 is also critical for the protein synthesis inhibitory effect of Shiga toxin in these cells. For example, toxin-sensitive lines such as HeLa or Vero cells always express Gb_3 (JACEWICZ et al. 1986, 1989; LINGWOOD et al. 1987; WADDELL et al. 1988), while refractory cells such as CHO cells or selected Daudi cell lines do not (COHEN et al. 1987; JACEWICZ et al. 1989; WADDELL et al. 1990). Among HeLa lines, the very sensitive HeLa 229 cells express significantly more Gb_3 than the more resistant CCL-2 cells (2.3×10^3 versus 0.4×10^3 pmol/mg. cell protein; JACEWICZ et al. 1989). Moreover, HeLa clones of increasing resistance to Shiga toxin obtained by a standard limiting dilution method show an inverse relation between sensitivity to toxin and the cell content of Gb_3. Inhibition of the HeLa cell cytotoxicity occurs when these cells are grown in the presence of tunicamycin, and this results in a concomitant decrease in the content of Gb_3 (JACEWICZ et al. 1989). This suggests the possibility that the high-affinity site in tunicamycin-treated cells might be Gb_3 rather than a glycoprotein as might have been expected. The mechanism by which tunicamycin affects Gb_3 content is not clear, but could indicate that tunicamycin-sensitive glycoproteins regulate Gb_3 levels in HeLa cells.

Attempts have also been made to insert Gb_3 into the plasma membrane of cells lacking Gb_3 and then to determine if this has an effect on their sensitivity to Shiga toxin. Recently, WADDELL et al. (1990) reported the results of experiments with toxin-resistant mutant Daudi cells incubated in the presence of liposomes made with phosphatidylethanolamine, phosphatidylserine and containing selected glycolipids. This resulted in the appearance of toxin-binding sites as Gb_3 liposomes became incorporated into the plasma membrane of the Daudi cells, and cytotoxic effects of toxin could be detected as inhibition of protein synthesis in the presence of the toxin. In a previous study, lecithin:cholesterol micelles containing Gb_3 were not incorporated into HeLa cell membranes, but did serve as a competitive inhibitor of toxin activity when added to the medium (JACEWICZ et al. 1986). Other glycolipids lacking the galα1-4gal disaccharide did not exhibit this competitive effect. Another approach has been to increase endogenous Gb_3 in cells by growing toxin-resistant HeLa clones lacking Gb_3 in the presence of 1,5-dideoxy-1,5-immino-D-galactitol (DIG), a competitive inhibitor of α-galactosidase A which prevents breakdown of Gb_3 to lactosylceramide. DIG-treated cells became sensitive to toxin in parallel with a significant increase in Gb_3 content and the appearance of toxin-binding sites (JACEWICZ et al. 1991).

The binding preference of other members of the Shiga toxin family differs somewhat from Shiga toxin itself or the very closely related SLT-I. Differential cytotoxic activities on HeLa and Vero cells is one of the main differentiating features of SLT-II and SLT-IIv (MARQUES et al. 1987). WEINSTEIN et al. (1989) were able to recombine the A and B subunit genes of Shiga toxin, SLT-I, SLT-II, and SLT-IIv in such a way that hybrids were formed in vivo. No important difference in HeLa and Vero cell cytotoxicity was found when any of the A subunits was combined with the B subunits from Shiga toxin or SLT-I or SLT-II. However, when the SLT-IIv B subunit was present, regardless of the source of the A subunit, the binding specificity of the hybird toxins was for Vero cells, with no activity to HeLa cells. DE GRANDIS et al. (1989) reported that SLT-IIv from a pig isolate (SLT-IIvp) bound poorly to Gb_3 on thin-layer chromatography (TLC) plates compared with SLT-I or SLT-II. The SLT-IIvp toxin demonstrated specific binding to globotetraosylceramide (Gb_4), a neutral glycolipid with terminal galNAcβ1-4 covering the penultimate galα1-4gal disaccharide. SAMUEL et al. (1990) subsequently compared binding specificity of SLT-IIvh from a human isolate and SLT-IIvp with SLT-II. *E. coli* strains carrying various fusions between SLT-II and SLT-IIv genes for the A and the B subunits were made in order to produce hybrid SLT-II/IIv toxins. As before, cell specificity was determined by the B subunit, independent of the source of the A subunit. Of interest in this study is that the glycolipid binding pattern for SLT-IIvp and SLT-IIvh differed between HeLa and Vero cells. Thus SLT-IIvp bound to Gb_3, Gb_4, and Gb_5 in

Vero cells, whereas SLT-IIvh bound only to Gb_3. This suggests that Gb_3, the common binding glycolipid for the SLT-IIv toxins, is the functional receptor in Vero cells, although the preferential binding of SLT-IIvp is to Gb_4. Even if this is true, binding to HeLa cell Gb_3 is not sufficient to result in the cytotoxicity. It should be clear from these data that the understanding of Shiga family toxin receptors remains incomplete at this time, and the important distinction between binding sites and receptors is again reinforced.

6 Endocytosis/Entry of Shiga Toxin

There is now good evidence that the enzymatically active Shiga toxin A chain enters the cytosol only after endocytic uptake of the toxin (SANDVIG et al. 1989a, 1991a, b). Although Shiga toxin is bound to sensitive cells by glycolipid receptors (COHEN et al. 1987; JACEWICZ et al. 1986; LINDBERG et al. 1987; MOBASSALEH et al. 1988), the toxin seems to be internalized from clathrin-coated pits (SANDVIG et al. 1989a, 1991a). When added to cells at 0 °C, the toxin is evenly bound at the cell surface, whereas after a short incubation at 37 °C the toxin is aggregated in clathrin-coated pits and is internalized by endocytosis (SANDVIG et al. 1989a). This has been shown both by using horse-radish-peroxidase-labeled toxin and by localization of the toxin with anti-Shiga toxin antiserum and protein G-gold (SANDVIG et al. 1989a). In fact, Shiga toxin seems to be the first example of a lipid-binding toxin that is internalized by this pathway. However, the mechanism behind the toxin-induced aggregation of glycolipid receptors in coated pits is not known. Other receptors known to use this pathway for internalization are dependent on a cytosolic protein tail (DAVIS et al. 1987; PRYWES et al. 1986; ROTHENBERGER et al. 1987). In contrast to Shiga toxin, the two glycolipid-binding toxins, cholera and tetanus toxin, are internalized from non-clathrin-coated areas of the cell membrane (MONTESANO et al. 1982; TRAN et al. 1987). Biochemical data support the results obtained with electron microscopy. The cells are protected against the toxin when the coated pit/coated vesicle pathway is blocked by acidification of the cytosol (HEUSER 1989; SANDVIG et al. 1987, 1988, 1989b), and the internalization process is rapid; the toxin is, in few minutes, inaccessible to antitoxin (KEUSCH 1981). Such rapid endocytosis is characteristic for internalization by clathrin-coated vesicles (for review see VAN DEURS et al. 1989).

Shortly after uptake, endocytosed ligands are exposed to low endosomal pH. However, in contrast to a number of other toxins (GORDON et al. 1988; OLSNES and SANDVIG 1988); Shiga toxin entry into the cytosol does not require low intravesicular pH. Addition of the ionophore nigericin, which will increase the pH in compartments that were originally acidified, does not protect against Shiga toxin (SANDVIG and BROWN, 1987). It appears that Shiga toxin has to be transported to a later compartment than the early endosome before translocation of the A chain can take place. The toxicity is strongly reduced when the temperature is lowered to 18 °C (SANDVIG et al. 1989a), a temperature at which fusion of endosomes with later compartments is inhibited (DUNN et al. 1980; VAN DEURS et al. 1987). Shiga toxin is transferred both to lysosomes and to the Golgi apparatus (SANDVIG et al.1989a, 1991a), but the experimental evidence suggests that it is the transport to the Golgi apparatus which is essential for the intoxication of the cells (SANDVIG et al. 1986, 1991b). Inhibitors of glycosylation and protein synthesis sensitize cells to Shiga toxin (SANDVIG et al. 1986), emphasizing the importance of the biosynthetic pathway (the Golgi apparatus) for the intoxication with Shiga toxin. Furthermore, the drug brefeldin A, which in a number of cell types leads to disintegration of most of the Golgi apparatus (DONALDSON et al. 1990; LIPPINCOTT-SCHWARTZ et al. 1989; SHITE et al. 1990; TAKAMI et al. 1990), protects these cells against Shiga toxin (SANDVIG et al. 1991b). This last result strongly suggests that Shiga toxin has to be transported to the Golgi apparatus to intoxicate cells.

The intracellular transport of Shiga toxin in a polarized epithelium has recently been studied (SANDVIG et al. 1991a). Polarized Madine Darby canine kidney (MDCK) cells grown on permeable filters originally lack binding sites for Shiga toxin and they are completely resistant. However, it turns out that upon incubation with butyric acid, glycolipids that bind Shiga toxin are synthesized and found in the cell membrane, and the cells do bind toxin and become sensitive. By use of electron microscopy and subcellular fractionation, it was shown that Shiga toxin can enter the Golgi apparatus both when added at the apical and the basolateral side of the cells, and quantitative studies showed that approximately 10% of the internalized Shiga toxin was transported to the Golgi apparatus per hour, the transport being equally efficient from the two poles of the epithelial cells. In agreement with that, the toxin was found to intoxicate the cells from both sides of the epithelial cell layer.

The details in the mechanism of transfer of the A chain across the membrane is not known. We know that Ca^{2+} is required and that inhibitors of Ca^{2+} transport inhibit the intoxication (SANDVIG and BROWN 1987), but we do not know which stage in the intoxication process is affected under these conditions. It is possible that it is the translocation step, as lack of Ca^{2+} and inhibitors of Ca^{2+} transport also inhibit the action of some plant toxins (SANDVIG and OLSNES 1982) which may enter the cytosol by using a similar mechanism as Shiga toxin.

7 Mode of Action of Shiga Toxin

Shiga toxin and SLTs act by inhibiting protein synthesis. The first indication that *S. dysenteriae* type 1 toxin inhibits protein synthesis in a cell-free system was obtained by THOMPSON et al. (1976). They observed that a partially purified toxin inhibited the poly-U-directed synthesis of polyphenylalanine. Later BROWN et al. (1980) showed that highly purified toxin inhibited protein synthesis in a rabbit reticulocyte lysate and that the inhibitory effect could be strongly increased by treatment of the toxin with proteolytic enzymes and reducing agents. Structural studies explained this activation. The inhibitory action on protein synthesis was found to be associated with the A chain of the toxin which is easily split by trypsin into a 27-kDa and a 4-kDa fragment bridged by a disulfide bond (OLSNES and EIKLID 1980; OLSNES et al. 1981). Although the whole toxin is necessary for toxic effect on whole cells, the larger part of the A chain, the A_1 fragment, inhibits cell-free protein synthesis. The activity of the free A_1 fragment was found to be much higher than that of whole toxin, and it was found to inactivate the 60*S* ribosomal subunits selectively (REISBIG et al. 1981).

BROWN et al. (1980) and REISBIG et al. (1981) noticed that each toxin molecule inactivated more than one ribosome, indicating catalytic activity. Since no cofactors had to be added, they suggested that the toxin would act as a hydrolytic enzyme. This suggestion was borne out by the discovery that the toxin A_1 fragment is a highly specific *N*-glycosidase that removes adenine from one particular adenosine residue in the 28*S* RNA of the 60*S* ribosomal subunit (ENDO et al. 1988; SAXENA et al. 1989). This particular adenosine residue is located on the vertex of a loop not far from the 3′ end of the 28*S* RNA (Fig. 2). In fact, Shiga toxin and SLTs were found to act in the same manner as ricin, abrin, and a number of related plant proteins. The A-chains of these toxins show a striking structure similarity with Shiga toxin and SLTs (CALDERWOOD et al. 1987; KOZLOV et al. 1987). Site-directed mutagenesis has implicated glutamic acid 167 in the active site of the toxin (HOVDE et al. 1988).

The toxins were found to be much more active on whole ribosomes than on the isolated 28*S* RNA, but also the latter was modified when toxin was

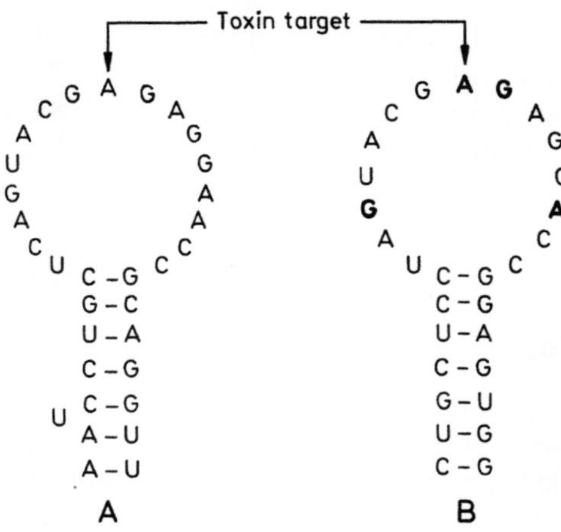

Fig. 2. Schematic structure of the region of rat liver 28-S RNA (*A*) and *E. coli* 23*S* RNA (*B*) that can act as toxin target. The residues indicated in bold phase in *B* represent contact sites for the bacterial elongation factors. In rat liver 28*S* RNA the target adenosine is in position 4324 from the 5′ end

present in high concentrations. Also, bacterial ribosomes were modified at high toxin concentrations (ENDO and TSURUGI 1988). The fact that eukaryotic ribosomes are much more sensitive than the isolated RNA indicates that the toxin recognizes additional structures in the intact ribosome.

It was known from earlier studies that ribosomes treated with ricin or Shiga toxin are unable to interact properly with elongation factors (FERNANDEZ-PUENTES et al. 1976; OBRIG et al. 1987; OLSNES et al. 1975). Recent studies on *E. coli* ribosomes have shown that a stem-loop structure of the 23*S* RNA which is related to the toxin target in eukaryotic ribosomes (Fig. 2) is involved in binding of elongation factors (MOAZED et al. 1988). As shown by ENDO et al. (1988) bacterial ribosomes are attacked by the toxins, although at a much lower rate than eukaryotic ribosomes.

8 Structure-Function Analysis of Shiga Toxin and the SLTs

8.1 Mutational Analysis of the Enzymatic Subunits

All members of the Shiga toxin family and ricin share two regions of amino acid sequence homology between residues 167–171 and 203–207 of the mature Shiga toxin A subunit (CALDERWOOD et al. 1987; DEGRANDIS et al. 1987; HOVDE et al. 1988; WEINSTEIN et al. 1988b). Because ricin, Shiga toxin, and the SLTs exert the same mechanism of action on eukaryotic ribosomes and these two homologous sequences lie within a putative active site cleft of the ricin A chain (MONTFORT et al. 1987), structure-function analysis of the enzymatic polypeptides has focused on these two conserved regions (JACKSON 1990). HOVDE et al. (1988) speculated that the glutamic acid residue at position 167 of the Shiga toxin/SLT-I A subunit may lie within a catalytic site of that polypeptide. This hypothesis was supported by previous findings in which carboxylate side chains were implicated as catalytic sites in glycosyl hydrolases and transferases, and that a glutamic acid residue was at the active sites of diphtheria toxin and *Pseudomonas aeruginosa* exotoxin A. The conservative substitution of glutamic acid 167 with an aspartic acid in the A subunits of Shiga toxin/SLT-I (HOVDE et al. 1988) and SLT-II (JACKSON et al. 1990a) resulted in a significant decrease in enzymatic activity, indicating that this residue is within an active site. Mutational analysis of the corresponding residue in the ricin A chain supported the role of this glutamic acid in catalysis, but also demonstrated that a neighboring arginine and the downstream region of sequence homology are required for activity (FRANKEL et al. 1989; SCHLOSSMAN et al. 1989). Based on these findings, FRANKEL et al. (1989) proposed that the downstream region of sequence homology between ricin and the Shiga toxins is required for 28*S* rRNA binding, while the upstream glutamic acid and arginine residues are required for the depurination reaction.

8.2 Mutational Analysis of the Binding Subunits

Mutational analysis of the Shiga toxin, SLT-II, and SLT-IIv B subunits identified residues which are required for receptor binding, recognition by neutralizing monoclonal antibodies, and extracellular localization in *E. coli* (JACKSON et al. 1990a, b; PERERA et al. 1991). However, unlike structure-function analysis of the enzymatic subunit, mutagenesis of the B subunit did not implicate a single active site residue. Instead, amino acid substitutions which reduced cytotoxic activity probably blocked receptor binding due to steric hindrance or by inducing a significant conformational change in the B polypeptides. One exception is the effect on extracellular localization by the relatively conservative change of glutamine to glutamic acid near the carboxyl terminus of SLT-IIvB (JACKSON et al. 1990b). This mutation dramatically reduced the B subunit-directed release of SLT-IIv from *E. coli*, suggesting that the introduction of a charged residue may have promoted ionic interactions between the holotoxin and the bacterial cell envelope.

In conclusion, members of the Shiga toxin family conform to the A/B structure found in other bacterial cytotoxins, although the enzymatic activity and receptor-binding specificity of the individual toxin subunits differ. The observation that Shiga toxin, the SLTs, and ricin inhibit eukaryotic protein synthesis by the identical mechanism of action directed the structure-function analysis of these glycosidases to two highly conserved regions in the otherwise divergent enzymatic subunits. The ricin A chain model proposes that these two regions are involved in the separate functions of 28*S* rRNA binding and single site depurination. While mutational analysis of the Shiga toxin and SLT A subunits supports the ricin model, verification will depend upon a comparison of the X-ray crystallographic structures of these molecules. Analysis of Shiga toxin and SLT hybrids revealed that the B subunits are required for receptor binding, cytotoxic specificity, and extracellular localization in *E. coli.* Mutational analysis suggests that residues which are distantly separated in the folded B polypeptide may be required for receptor binding. However, identification of B subunit-binding sites may ultimately depend upon the analysis of purified receptor-ligand complexes.

9 Role of Shiga Toxin and the SLTs in Pathogenesis

The role of Shiga toxin in the pathogenesis of shigellosis has been the subject of debate virtually from the time of its initial description until today (KEUSCH et al. 1986a). In contrast, the importance of Shiga and Shiga family toxins in extraintestinal manifestations of infection has been much less controversial. It is ironic that the problem initially experienced in separating the effects of the protein Shiga toxin from lipopolysaccharide (LPS) endotoxin in the first 50 years of its study and overcome in recent years because highly purified toxin with minimal contamination by LPS was available, as well as the use of assays insensitive to LPS (the rabbit ileal loop and HeLa or Vero lethal cytotoxicity assays), may return to plague interpretation of physiological studies using endothelial cells, which can be activated to a proinflammatory and procoagulant state by miniscule quantities of LPS.

Until 1972, most studies with Shiga toxin entailed parenteral injections of endotoxin-contaminated toxin preparations and use of the "neurotoxin" (limb paralysis/lethality) assay. With the application of more relevant intestinal models such as the ligated rabbit ileal loop, Shiga toxin was shown to be "enterotoxic," resulting in net accumulation of fluid in toxin-exposed gut loops (KEUSCH et al. 1972a) and associated with an inflammatory enteritis (KEUSCH et al. 1972b). In the two decades since then, the role of toxin in pathogenesis of intestinal manifestations of shigellosis has neither been confirmed nor disproved. There is no doubt, however, about the ability of toxin to cause fluid accumulation in the rabbit small bowel. This is secondary to reduced sodium absorption, attributed to villus cell dysfunction resulting from toxin targeting to surface expressed Gb_3 on villus, not crypt, cells (KANDEL et al. 1989).

There is ample evidence that Shiga toxin exerts cytotoxic effects on intestinal epithelial cells, including human colonic cells in primary culture (MOYER et al. 1987). However, the in vivo situation is much more complicated since not only is there free toxin in the lumen, but shigellae also invade and multiply within epithelial cells (SANSONETTI 1991), making it difficult to readily distinguish between invasion and an elicited inflammatory response and the specific direct effects of toxin.

Both a chlorate-resistant mutant lacking the Shiga toxin gene (LEVINE et al. 1973; NEILL et al. 1988) and a specific toxin deletion mutant of *S. dysenteriae* type 1 (FONTAINE et al. 1988) retain the ability to produce intestinal disease in primates (including humans), albeit less severe than disease caused by the wild-type Tox^+ strain. However, both mutants produce low levels of a cytotoxin under certain conditions in vitro, and human volunteers infected with the chlorate-resistant strain produce serum antibody cross-reactive with Shiga toxin. Thus, these studies do not exclude the possibility that Shiga family toxins play a role in causing the gut pathology of shigellosis. In fact, the presence of the functional *stx* gene was associated with severe inflammatory lesions of the colon in rhesus monkey infections, not seen with the Tox^- mutant (FONTAINE et al. 1988). These lesions were characterized by destruction of the capillaries serving the colonic mucosa and an inflammatory vasculitis. Therefore, the major contributory role of the toxins in colonic disease may be in mediating vascular damage. The importance of this observation should not be underestimated since the disruption of the integrity of the microvessels serving the colon may increase the access of the toxins and other bacterial products to the bloodstream and increase the likelihood of developing serious postdysenteric sequelae.

Numerous epidemiological studies of outbreaks of hemorrhagic colitis and the hemolytic uremic syndrome (HUS) have definitively established an association between the disorders and infection with SLT-producing *E. coli* (BOPP et al. 1987; CARTER et al. 1987; KARMALI et al. 1983, 1985; KLEANTHOUS et al. 1990; OSTROFF et al. 1989; REMIS et al. 1984). In addition, isolation of SLT-I- and/or SLT-II-producing *E. coli* is occasionally associated with thrombotic thrombocytopenic purpura. Early epidemiological studies clearly demonstrated that isolation of *E. coli* serotype O157:H7, a hitherto "rare" serotype, was associated with severe outbreaks of bloody diarrhea in the United States and Canada (JOHNSON et al. 1983; O'BRIEN et al. 1983; RILEY et al. 1983; WELLS et al. 1983). Strains of *E. coli* producing SLTs have thus been categorized as enterohemorrhagic *E. coli* or EHEC (LEVINE 1987). Unlike *S. dysenteriae* type 1, EHEC are not invasive, and the capacity of EHEC strains to adhere to colonic epithelia and colonize the gut may be a critical virulence determinant. The precise role of the SLTs in the pathogenesis of hemorrhagic colitis and HUS is, at present, not fully understood.

The best evidence in favor of an effect of the toxin on the gut in vivo has been the ability of noninvasive EHEC O157:H7 (DONNENBERG et al. 1989), which produce large amounts of SLT-I and/or SLT-II, to cause colonic damage which ranges from inflammatory changes with or without ulcerations to pseudomembranous colitis, and occasionally to exacerbations of ulcerative colitis (HUNT et al. 1989; LJUNGH et al. 1988; VON WULFEN et al. 1989). Hemorrhage and edema are found in the lamina propria of the colon, with focal necrosis superficially but preservation of the deeper crypts, which resembles the pathology of colitis due to ischemia or *Clostridium difficile* (GRIFFIN et al. 1990).

The discovery of the *eae* gene, which mediates the attaching and effacing lesion caused by EHEC and class I enteropathogenic *E. coli* (EPEC), indicates that MVM may also be directly damaged by the *eae* gene product (JERSE et al. 1990), independent of any Shiga family toxin effect. Nonetheless, there is a significant difference in the severity of the pathology and the nature of the clinical disease caused by EPEC strains expressing the *eae* gene but producing only low levels of cytotoxin and the EHEC strains which are eae^{+} and make large quantities of the toxin. The latter cause marked changes in morphology and result in bloody diarrhea or hemorrhagic colitis.

No entirely satisfactory small animal model of hemorrhagic colitis or HUS is currently available, although many attempts to develop one have been made. The morphological effects of Shiga toxin and SLT from *E. coli* O157:H7 have been compared in the rabbit intestine, and the effects of the two toxins are similar (KEENAN et al. 1986), consisting of dose-dependent villous blunting and decrease in the villus/crypt ratio. BARRETT et al. (1989a) placed an osmotic pump in the rabbit peritoneal cavity for continuous infusion of SLT-II and induced diarrhea and hemorrhagic gut lesions, although none of the animals developed renal failure. When LPS and SLT-II were given together, there was enhanced lethality, although pretreatment with LPS protected the animals from the lethal effect of SLT-II challenge (BARRETT et al. 1989b).

PAI et al. (1986) reported that rabbits infected with toxin-positive *E. coli* O157:H7, but not toxin-negative O157:H45, developed bloody diarrhea with histological evidence of mucosal damage; they correlated this with the presence of free toxin in the mid-colon. Similar results were obtained by giving toxin alone, further suggesting its importance in pathogenesis. WADOLKOWSKI et al. (1990a) infected streptomycin-treated mice with *E. coli* O157:H7, which preferentially colonized the gut compared to an isogenic strain cured of the 60 MDa plasmid encoding attachment fimbriae. However, none of the animals developed diarrhea. In vivo passage of the plasmid-cured strain led to the isolation of a colonizing variant able to cause a fatal illness 4–10 days postfeeding, manifested by lethargy and anorexia with either loose stools or constipation. While these animals had widespread, bilateral, severe acute cortical necrosis, no lesions were found in the colon. When the animals were pretreated with antibodies to SLT-I and SLT-II, either alone or in combination, or when they were infected with an *E. coli* K-12 strain expressing cloned SLT-I or SLT-II genes, renal lesions and death were noted only when SLT-II was present (WADOLKOWSKI et al. 1990b). Although there was little pathological similarity between the tubular lesion found in the mice and the glomerular lesions in humans with HUS, the association of the former with EHEC-producing SLT-II is consistent with epidemiological evidence which suggests that HUS in humans is particularly associated with SLT-II-producing EHEC as well (KLEANTHOUS et al. 1990; MILFORD et al. 1990; OSTROFF et al. 1989).

The association of HUS and Shiga family toxins is further supported by the epidemiological linkage of cancer-associated HUS with clinical use of mitomycin C for chemotherapy (LESESNE et al. 1989). This drug dramatically increases the levels of SLT-I and especially SLT-II produced in vitro because it increases the replication of the transducing bacteriophage-carrying toxin genes (ACHESON et al. 1990), and it is possible that clinical use of mitomycin C during cancer could induce increased SLT production from *E. coli* present in the gut flora sufficient to cause damage to glomerular endothelium and initiate HUS (ACHESON and DONOHUE-ROLFE 1989).

The collective results of numerous histopathological studies of human patients with bacillary dysentery, hemorrhagic colitis, or HUS and the animal studies cited above have led to the concept that the common characteristic of both the prodromal diarrheal disease and the sequelae may be toxin-mediated vascular damage in the target organs (CLEARY and LOPEZ 1989; KAVI and WISE 1989; MILFORD and TAYLOR 1990; OBRIG et al. 1988; TESH and O'BRIEN 1991). Although originally described as a functionally quiescent partition between blood and the interstitium, it is now clear that vascular endothelial cells are critical for the maintenance of transcapillary permeability and the nonthrombogenic state necessary for normal blood flow. Direct toxin-mediated endothelial cell damage may not only lead to a procoagulant state, but also result in the elicitation of endogenous cytokines that, in turn, may exacerbate cell damage (TESH et al. 1991), further alter hemostasis and vascular permeability (BEVILACQUA et al. 1984, 1986; NACHMAN et al. 1986; NAWROTH and STERN 1986; ROYALL et al. 1989; VAN HINSBERGH et al. 1988), and mediate the expression of membrane receptors necessary for monocyte- and leukocyte-endothelial cell adherence (BEVILACQUA et al. 1985; CAVENDER et al. 1986; GAMBLE et al. 1985; SCHLEIMER and RUTLEDGE 1986). In addition, Shiga toxin and SLTs may act directly on endothelial cells to inhibit the production of prostacyclin (KARCH et al. 1988), and this may result in the induction of the platelet-aggregating activity characteristic of plasma from HUS patients (ROSE et al. 1985).

Another association of Shiga family toxins and clinical manifestations is the role of SLT-IIv toxin in pathogenesis of edema disease in piglets. This disease in swine is characterized by edema of subcutaneous tissues, stomach wall, mesentery, and other tissues, neurological manifestations such as incoordination and staggering gait, limb paralysis and death, and is associated with certain serotypes of *E. coli*, now known to produce SLT-IIv toxins (GYLES et al. 1988; WEINSTEIN et al. 1988b). These phenomena appear to be a direct consequence of the toxin, as they can be reproduced by injection of purified SLT-IIv (MACLEOD and GYLES 1990) and because toxin-negative strains fail to cause edema or cerebral lesions (FRANCIS et al. 1989). The relevance of this syndrome to any of the manifestations of human shigellosis or *E. coli* SLT- induced disease is, at present, uncertain.

References

Acheson DWK, Donohue-Rolfe A (1989) Cancer associated hemolytic uremic syndrome. A possible role of mitomycin C in relation to Shiga-like toxins. J Clin Oncol 7: 1943

Acheson DWK, Muhldorfer I, Kane A, Keusch GT, Donohue-Rolfe A (1990) Bacteriophage induction as a cause of increased Shiga-like toxin synthesis in *E. coli.* Abstracts of the Annual meeting of the American Society for Microbiology Washington, DC, p 61

Barrett TJ, Potter ME, Wachsmuth IK (1989a) Continuous peritoneal infusion of Shiga-like toxin II (SLT II) as a model for SLT II-induced diseases. J Infect Dis 159: 774–777

Barrett TJ, Potter ME, Wachsmuth IK (1989b) Bacterial endotoxin both enhances and inhibits the toxicity of Shiga-like toxin II in rabbits and mice. Infect Immun 57: 3434–3437

Betley MJ, Miller VL, Mekalanos JJ (1986) Genetics of bacterial enterotoxins. Annu Rev Microbiol 40: 577–605

Bevilacqua MP, Pober JS, Majeau GR, Cotran RS, Gimbrone MA Jr (1984) Interleukin (IL-1) induces biosynthesis and cell surface expression of procoagulant activity in human vascular endothelial cells. J Exp Med 160: 618–623

Bevilacqua MP, Pober JS, Wheeler ME, Cotran RS, Gimbrone MA Jr (1985) Interleukin-1 acts on cultured human vascular endothelial cells to increase the adhesion of polymorphonuclear leukocytes, monocytes, and related cell lines. J Clin Invest 76: 2003–2011

Bevilacqua MP, Pober JS, Majeau GR, Fiers W, Cotran RS, Gimbrone MA Jr (1986) Recombinant tumor necrosis factor induces procoagulant activity in cultured human vascular endothelium: characterization and comparison with the action of interleukin 1. Proc Natl Acad Sci USA 83: 4533–4537

Bopp CA, Greene KD, Downes F, Sowers EG, Wells JG, Wachsmuth IK (1987) Unusual verotoxin-producing *Escherichia coli* associated with hemorrhagic colitis. J Clin Microbiol 25: 1485–1489

Bridgwater FAJ, Morgan RS, Rowson KEK, Payling-Wright G (1955) The neurotoxin of *Shigella shigae*. Morphological and functional lesions produced in the central nervous system of rabbits. Br J Exp Pathol 36: 447–453

Brown JE, Ussery MA, Leppla SH, Rothman SW (1980) Inhibition of protein synthesis by Shiga toxin. Activation of toxin and inhibition of peptide elongation. FEBS Lett 117: 84–88

Brown JE, Griffin DE, Rothman SW, Doctor BP (1982) Purification and biological characterization of Shiga toxin for *Shigella dysenteriae*. Infect Immun 36: 996–1005

Brown JE, Karlsson K-A, Lindberg A, Stromberg N, Thurin J (1983) Identification of the receptor glycolipid for the toxin of *Shigella dysenteriae*. In: Chester MA, Heinegaard D, Lundblad A, Svensson S (eds.) Proceedings of the 7th International Symposium on Glycoconjugates. Rahms i Lund, Lund, Sweden, pp 678–679

Calderwood SB, Mekalanos JJ (1987) Iron regulation of Shiga-like toxin expression in *Escherichia coli* is mediated by the *fur* locus. J Bacteriol 169: 4759–4764

Calderwood SB, Mekalanos JJ (1988) Confirmation of the Fur operator site by insertion of a synthetic oligonucleotide into an operon fusion plasmid. J Bacteriol 170: 1015–1017

Calderwood SB, AuClair F, Donohue-Rolfe A, Keusch GT, Mekalanos JJ (1987) Nucleotide sequence of the Shiga-like toxin genes of *Escherichia coli*. Proc Natl Acad Sci USA 84: 4364–4368

Calderwood SB, Acheson DWK, Goldberg MB, Boyko SA, Donohue-Rolfe A (1990) A system of production and rapid purification of large amounts of the Shiga toxin/Shiga-like toxin I B subunit. Infect Immun 58: 2977–2982

Carter AO, Borczyk AA, Carlson JAK, Harvey B, Hockin JC, Karmali MA, Krishnan C, Kron DA, Lior H (1987) A severe outbreak of *Escherichia coli* O157: H7-associated hemorrhagic colitis in a nursing home. N Engl J Med 317: 1496–1500

Cavender DE, Haskard DO, Joseph B, Ziff M (1986) Interleukin 1 increases the binding of human B and T lymphocytes to endothelial cell monolayers. J Immunol 136: 203–207

Chantemesse A, Widal F (1888) Sur les microbes de la dysenteria épidémique. Bull Acad Med (Paris) 19: 522–529

Cleary TG, Lopez EL (1989) The Shiga-like toxin-producing *Escherichia coli* and hemolytic uremic syndrome. Pediatr Infect Dis J 8: 720–724

Cohen A, Hannigan GE, Williams BRG, Lingwood CA (1987) Roles of globotriosyl- and galabiosylceramide in Verotoxin binding and high affinity interferon receptor. J Biol Chem 262: 17088–17091

Conradi H (1903) Über lösliche, durch aseptische Autolyse erhaltene Giftstoffe von Ruhr- und Typhus-Bazillen. Dtsch Med Wochenschr 29: 26–28

Davis CG, van Driel IR, Russell DW, Brown MS, Goldstein JL (1987) The low density lipoprotein receptor. Identification of amino acids in cytoplasmic domain required for rapid endocytosis. J Biol Chem 262: 4075–4082

DeGrandis S, Ginsberg J, Toone M, Climie S, Friesen J, Brunton J (1987) Nucleotide sequence and promoter mapping of the *Escherichia coli* Shiga-like toxin operon of bacteriophage H-19B. J Bacteriol 169: 4313–4319

DeGrandis S, Law H, Brunton J, Gyles C, Lingwood CA (1989) Globotetraosyl ceramide is recognized by the pig edema disease toxin. J Biol Chem 264: 12520–12525

Donaldson JG, Lippincott-Schwartz J, Bloom GS, Kreis TE, Klausner RD (1990) Dissociation of a 110-kD peripheral membrane protein from the Golgi apparatus is an early event in brefeldin A action. J Cell Biol 111: 2295–2306

Donnenberg MS, Donohue-Rolfe A, Kesch GT (1989) Epithelial cell invasion: an overlooked property of enteropathogenic *Escherichia coli* (EPEC) associated with the EPEC adherence factor. J Infect Dis 160: 452–459

Donohue-Rolfe A, Keusch GT (1983) *Shigella dysenteriae* 1 cytotoxin: periplasmic protein releasable by polymyxin B and osmotic shock. Infect Immun 39: 270–274

Donohue-Rolfe A, Keusch GT, Edson C, Thorley-Lawson D, Jacewicz M (1984) Pathogenesis of shigella diarrhea. IX. Simplified high yield purification of shigella toxin and characterization of subunit composition and function by the use of subunit-specific monoclonal and polyclonal antibodies. J Exp Med 160: 1767–1781

Donohue-Rolfe A, Acheson DWK, Kane AV, Keusch GT (1989a) Purification of Shiga toxin and Shiga-like toxins I and II by receptor analog affinity chromatography with immobilized P1 glycoprotein and production of cross-reactive monoclonal antibodies. Infect Immun 56: 3888–3893

Donohue-Rolfe A, Jacewicz M, Keusch GT (1989b) Isolation and characterization of functional Shiga toxin subunits and renatured holotoxins. Mol Microbiol 3: 1231–1236

Dubos RJ, Geiger JW (1946) Preparation and properties of Shiga toxin and toxoid. J Exp Med 84: 143–156

Dunn WA, Hubbard AL, Aronson NN Jr (1980) Low temperature selectively inhibits fusion between pinocytic vesicles and lysosomes during heterophagy of ^{125}I-asialofetuin by the perfused rat liver. J Biol Chem 255: 5971–5978

Endo Y, Tsurugi K (1988) The RNA N-glycosidase activity of ricin A-chain. The characteristics of the enzymatic activity of ricin A-chain with ribosomes and with rRNA. J Biol Chem 263: 8735–8739

Endo Y, Tsurugi K, Yutsudo T, Takeda Y, Ogasawara T, Igarashi K (1988) Site of action of Vero toxin (VT2) from *Escherichia coli* O157: H7 and of Shiga toxin on eukaryotic ribosomes. RNA N-glycosidase activity of the toxins. Eur J Biochem 171: 45–50

Fernandez-Puentes C, Benson S, Olsnes S, Pihl A (1976) Protective effect of elongation factor 2 on the inactivation of ribosomes by the toxic lectins abrin and ricin. Eur J Biochem 64: 437–443

Flexner S (1990) On the etiology of tropical dysentery. Bull Johns Hopkins Hosp 11: 231–242

Fontaine A, Arondel J, Sansonetti PJ (1988) Role of Shiga toxin in the pathogenesis of bacillary dysentery studied using a tox$^-$ mutant of *Shigella dysenteriae* 1. Infect Immun 56: 3099–3109

Francis DH, Moxley RA, Andraos CY (1989) Edema disease-like brain lesions in gnotobiotic piglets infected with *Escherichia coli* serotype O157: H7. Infect Immun 57: 1339–1342

Frankel A, Schlossman D, Welsh P, Hertler A, Withers D, Johnston S (1989) Selection and characterization of ricin toxin A-chain mutations in *Saccharomyces cerevisiae.* Mol Cell Biol 9: 415–420

Fuchs G, Mobassaleh M, Donohue-Rolfe A, Montgomery RK, Grand RJ, Keusch GT (1986) Pathogenesis of *Shigella* diarrhea: rabbit intestinal cell microvillus membrane binding site for *Shigella* toxin. Infect Immun 53: 372–377

Gamble JR, Harlan JM, Klebanoff SJ, Vadas MA (1985) Stimulation of adherence of neutrophils to umbilical vein endotheliun by human recombinant tumor necrosis factor. Proc Natl Acad Sci USA 82: 8667–8671

Gannon VPJ, Teerling C, Masrei SA, Gyles CL (1990) Molecular cloning and nucleotide sequence of another variant of the *Escherichia coli* Shiga-like toxin II family. J Gen Microbiol 136: 1125–1135

Gemski P, Takeuchi A, Washington O, Formal SB (1972) Shigellosis due to *Shigella dysenteriae* 1: relative importance of mucosal invasion versus toxin production in pathogenesis. J Infect Dis 126: 523–530

Gill DM (1976) The arrangements of subunits in cholera toxin. Biochemistry 15: 1242–1248

Gordon VM, Leppla SH, Hewlett EL (1988) Inhibitors of receptor-mediated endocytosis block the entry of *Bacillus anthracis* adenylate cyclase toxin but not that of *Bordetella pertussis* adenylate cyclase toxin. Infect Immun 56: 1066–1069

Griffin PM, Olmstead LC, Petras RE (1990) *Escherichia coli* O157: H7-associated colitis. A clinical and histological study of 11 cases. Gastroenterology 99: 142–149

Gyles CL, DeGrandis SA, MacKenzie C, Brunton JL (1988) Cloning and nucleotide sequence analysis of the genes determining verocytotoxin production in a porcine edema disease isolate of *Escherichia coli.* Microb Pathog 5: 419–426

Hart PJ, Monzingo AF, Donohue-Rolfe A, Keusch GT, Calderwood SB, Robertus JD (1991) Crystallization of the B chain of Shiga-like toxin I from *Escherichia coli.* J Mol Biol 218: 691–694

Heuser J (1989) Changes in lysosome shape and distribution correlated with changes in cytoplasmic pH. J Cell Biol 108: 855–864

Hovde CJ, Calderwood SB, Mekalanos JJ, Collier RJ (1988) Evidence that glutamic acid 167 is an active-site residue of Shiga-like toxin I. Proc Natl Acad Sci USA 85: 2568–2572

Howard JG (1955) Observations on the intoxication produced in mice and rabbits by the neurotoxin of *Shigella shigae.* Br J Exp Pathol 36: 439–446

Huang A, DeGrandis S, Friesen J, Karmali M, Congi R, Brunton JL (1986) Cloning and expression of the genes specifying Shiga-like toxin production in *Escherichia coli* H19. J Bacteriol 166: 375–379

Huang A, Friesen J, Brunton JL (1987) Characterization of a bacteriophage that carries the genes for production of Shiga-like toxin 1 in *Escherichia coli.* J Bacteriol 169: 4308–4312

Hunt CM, Harvey JA, Youngs ER, Irwin ST, Reid TM (1989) Clinical and pathological variability of infection by enterohaemorrhagic (Vero cytotoxin producing) *Escherichia coli.* J Clin Pathol 42: 847–852

Ito H, Yutsudo T, Hirayama T, Takeda Y (1988) Isolation and some properties of A and B subunits of Vero toxin 2 and in vitro formation of hybrid toxins between subunits of Vero toxin 1 and Vero toxin 2 from *Escherichia coli* O157: H7. Microb Pathog 5: 189–195

Ito H, Terai A, Kurazono H, Takeda Y, Nishibuchi M (1990) Cloning and nucleotide sequencing of Vero toxin 2 variant genes from the *Escherichia coli* O91: H21 isolated from a patient with the hemolytic uremic syndrome. Microb Pathog 8: 47–60

Jacewicz M, Clausen H, Nudelman E, Donohue-Rolfe A, Keusch GT (1986) Pathogenesis of shigella diarrhea. XI. Isolation of shigella toxin-binding glycolipid from rabbit jejunum and HeLa cells and its identification as globotriaosyl-ceramide. J Exp Med 163: 1391–1404

Jacewicz M, Feldman HA, Donohue-Rolfe A, Balasubramanian KA, Keusch GT (1989) Pathogenesis of Shigella diarrhea. XIV. Analysis of Shiga toxin receptors on cloned HeLa cells. J Infect Dis 159: 881–889

Jacewicz MS, Gross SK, Mobassaleh M, Balasubramanian KA, Daniel P, McCluer RH, Keusch GT (1991) Pathogenesis of Shigella diarrhea. The relationship between Gb_3 content of HeLa cells and their sensitivity to Shiga toxin. (manuscript submitted)

Jackson MP (1990) Structure-function analyses of Shiga toxin and the Shiga-like toxins. Microb Pathog 8: 235–242

Jackson MP, Neill RJ, O'Brien AD, Holmes RK, Newland JW (1987a) Nucleotide sequence analysis and comparison of the structural genes for Shiga-like toxin I and Shiga-like toxin II encoded by bacteriophages from *Escherichia coli* 933. FEMS Lett 44: 109–114

Jackson MP, Newland JW, Holmes RK, O'Brien AD (1987b) Nucleotide sequence analysis of the structural genes for Shiga-like toxin I encoded by bacteriophage 933J from *Escherichia coli*. Microb Pathog 2: 147–153

Jackson MP, Deresiewicz RL, Calderwood SB (1990a) Mutational analysis of the Shiga toxin and Shiga-like toxin II enzymatic subunits. J Bacteriol 172: 3346–3350

Jackson MP, Wadolkowski EA, Weinstein DL, Holmes RK, O'Brien AD (1990b) Functional analysis of the Shiga toxin and Shiga-like toxin type II variant binding subunits using site-directed mutagenesis. J Bacteriol 172: 653–658

Jerse AE, Yu J, Tall BD, Kaper JH (1990) A genetic locus of enteropathogenic *Escherichia coli* necessary for the production of attaching and effacing lesions on tissue culture cells. Proc Natl Acad Sci USA 87: 7839–7843

Johnson WM, Lior H, Bezanson GS (1983) Cytotoxic *Escherichia coli* O157: H7 associated with haemorrhagic colitis in Canada. Lancet i: 76

Kandel G, Donohue-Rolfe A, Donowitz M, Keusch GT (1989) Pathogenesis of Shigella diarrhea. XVI. Selective targeting of Shiga toxin to villus cells of rabbit jejunum explains the effect of the toxin on intestinal electrolyte transport. J Clin Invest 84: 1509–1517

Karch H, Bitzan M, Pietsch R, Stenger K-O, von Wulffen H, Heesemann J, Dusing R (1988) Purified verotoxins of *Escherichia coli* O157: H7 decrease prostacyclin synthesis by endothelial cells. Microb Pathog 5: 215–221

Karmali MA, Steele BT, Petric M, Lim C (1983) Sporadic cases of hemolytic uremic syndrome associated with fecal cytotoxin and cytotoxin-producing *Escherichia coli*. Lancet i: 619–620

Karmali MA, Petric M, Lim C, Fleming PC, Arbus GS, Lior H (1985) The association between hemolytic uremic syndrome and infection by verotoxin-producing *Escherichia coli*. J Infect Dis 151: 775–782

Kavi J, Wise R (1989) Causes of the hemolytic uraemic syndrome. Br Med J 298: 65–66

Keenan KP, Sharpnack DD, Collins H, Formal SB, O'Brien AD (1986) Morphologic evaluation of the effects of Shiga toxin and *Escherichia coli* Shiga-like toxin on the rabbit intestine. Am J Pathol 125: 69–80

Keusch GT (1981) Receptor-mediated endocytosis of Shigella cytotoxin. In: Middlebrook JL, Kohn LD (eds) Receptor-mediated binding and internalization of toxins and hormones. Academic, New York; pp 95-112

Keusch GT, Jacewicz M (1977) Pathogenesis of *Shigella* diarrhea. VII. Evidence for a cell membrane toxin receptor involving β1-4 linked N-acetyl-D-glucosamine oligomers. J Exp Med 146: 535–546

Keusch GT, Grady GF, Mata LJ, McIver J (1972a) The pathogenesis of *Shigella* diarrhea. I. Enterotoxin production by *Shigella dysenteriae* 1. J Clin Invest 51: 1212–1218

Keusch GT, Grady GF, Takeuchi A, Sprinz H (1972b) The pathogenesis of *Shigella* diarrhea. II. Enterotoxin induced acute enteritis in the rabbit ileum. J Infect Dis 126: 92–95

Keusch GT, Donohue-Rolfe A, Jacewicz M (1986a) *Shigella* toxin(s): description and role in diarrhea and dysentery. In: Dorner F, Drews J (eds) Pharmacology of bacterial toxins. Pergamon, Oxford pp 235–270

Keusch GT, Jacewicz M, Donohue-Rolfe A (1986b) Pathogenesis of *Shigella* diarrhea: evidence for N-linked glycoprotein *Shigella* receptors and receptor modulation by β-galactosidase. J Infect Dis 153: 238–248

Keusch GT, Donohue-Rolfe A, Jacewicz M, Kane AV (1988) Shiga toxin: production and purification. In: Harshman S (ed) Methods in enzymology, vol 165. Microbial toxins: tools in enzymology. Academic, San Diego, 152(62): 399–401

Kleanthous H, Smith HR, Scotland SM, Gross RJ, Rowe B, Taylor CM, Milford DV (1990) Haemolytic uraemic syndromes in the British isles, 1985–8: association with Verocytotoxin-producing *Escherichia coli.* Part 2: microbiological aspects. Arch Dis Child 65: 722–727

Konowalchuk J, Speir JI, Stavric S (1977) Vero response to a cytotoxin of *Escherichia coli.* Infect Immun 18: 775–779

Konowalchuk J, Dickie N, Stavric S, Speir JI (1978) Comparative studies of five heat-labile toxic products of *Escherichia coli.* Infect Immun 22: 644–648

Kozlov YV, Kabishev AA, Fedchenko VI, Bayev AA (1987) Cloning and sequencing of Shiga-toxin structural genes. Proc Acad Sci USSR 295: 740–744

Kozlov YV, Kabishev AA, Lukyanov EV, Bayev AA (1988) The primary structure of operons coding for *Shigella dysenteriae* toxin and temperate phage H30 Shiga-like toxin. Gene 67: 213–221

Lesesne JB, Rothschild N, Erickson B, Korec S, Sisk R, Keller J, Arbus M, Woolley PV, Chiazze L, Schein PS, Neefe JR (1989) Cancer-associated hemolytic-uremic syndrome: an analysis of 85 cases from a national registry. J Clin Oncol 7: 781–789

Levine MM (1987) *Escherichia coli* that cause diarrhea: enterotoxigenic, enteropathogenic, enteroinvasive, enterohemorrhagic, and enteroadherent. J Infect Dis 155: 377–389

Levine MM, DuPont HL, Formal SB, Hornick RB, Takeuchi A, Gangarosa EJ, Snyder MJ Libonati JP (1973) Pathogenesis of *Shigella dysenteriae* 1 (Shiga) dysentery. J Infect Dis 127: 261–270

Lindberg AA, Brown JE, Stromberg N, Westling-Ryd M, Schultz JE, Karlsson KA (1987) Identification of the carbohydrate receptor for Shiga toxin produced by *Shigella dysenteriae* type 1. J. Biol Chem 262: 1779–1785

Linggood MA, Thompson JM (1987) Verotoxin production among porcine strains of *Escherichia coli* and its association with oedema disease. J Med Microbiol 25: 359–362

Lingwood CA, Law H, Richardson S, Petric M, Brunton JL, Karmali M (1987) Glycolipid binding of purified and recombinant *Escherichia coli* produced Verotoxin in vitro. J Biol Chem 262: 8834–8839

Lippincott-Schwartz J, Yuan LC, Bonifacino JS, Klausner RD (1989) Rapid redistribution of Golgi proteins into the ER in cells treated with brefeldin A: evidence for membrane cycling from Golgi to ER. Cell 56: 801–813

Ljungh A, Erickson M, Erickson O, Henter JI, Wadstrom T (1988) Shiga-like toxin production and connective tissue protein binding of *Escherichia coli* isolated from a patient with ulcerative colitis. Scand J Infect Dis 20: 443–446

Losch F (1875) Massenhafte Entwickelung von Aeroben im Dickdarm. Arch Pathol Anat Physiol 65: 196–211

MacLeod DL, Gyles CL (1990) Purification and characterization of an *Escherichia coli* Shiga-like toxin II variant. Infect Immun 58: 1232–1239

Marques LRM, Peiris JSM, Cryz SJ, O'Brien AD (1987) *Escherichia coli* strains isolated from pigs with edema disease produce a variant of Shiga-like toxin II. FEMS Lett 44: 33–38

Maurelli AT, Blackmon B, Curtiss R III (1984) Temperature-dependent expression of virulence genes in *Shigella* species. Infect Immun 43: 195–201

Milford DV, Taylor CM (1990) New insights into the hemolytic uraemic syndromes. Arch Dis Child 65: 713–715

Milford DV, Taylor CM, Guttridge B, Hall S, Rowe B, Kleanthous H (1990) Haemolytic uraemic syndromes in the British isles, 1985–8: association with Verocytotoxin producing *Escherichia coli.* Part 1: clinical and epidemiologic aspects. Arch Dis Child 65: 716–721

Moazed D, Robertson JM, Noller HF (1988) Interaction of elongation factors EF-G and EF-Tu with a conserved loop in 23S RNA. Nature 334: 362–364

Mobassaleh M, Donohue-Rolfe A, Jacewicz M, Grand RJ, Keusch GT (1988) Pathogenesis of Shigella diarrhea: evidence for a developmentally regulated glycolipid receptor for Shigella toxin involved in the fluid secretory response of rabbit small intestine. J Infect Dis 157: 1023–1031

Mobassaleh M, Gross SK, McCluer RH, Donohue-Rolfe A, Keusch GT (1989) Quantitation of the rabbit intestinal glycolipid receptor for Shiga toxin. Further evidence for the developmental regulation of globotriaosylceramide in microvillus membranes. Gastroenterology 97: 384–391

Mobassaleh M, Gross SK, McCluer RH, Keusch GT (1991) Distribution, subcellular localization and elucidation of the development regulation of Shiga toxin receptor synthesis in rabbit small intestine. Abstracts of the annual meeting of the American Gastroenterological Association, Washington

Montesano R, Roth J, Robert A, Orci L (1982) Non-coated membrane invaginations are involved in binding internalization of cholera and tetanus toxins. Nature 296: 651–653
Montfort W, Villafranca JE, Monzingo AF, Ernst SR, Katzin B, Rutenber E, Xuong NH, Hamlin R, Robertus JD (1987) The three-dimensional structure of ricin at 2.8Å. J Biol Chem 262: 5398–5403
Moyer MP, Dixon PS, Rothman SW, Brown JE (1987) Cytotoxicity of Shiga toxin for primary cultures of human colonic and ileal epithelial cells. Infect Immun 55: 1533–1535
Nachman RL, Hajjar KA, Silverstein RL, Dinarello CA (1986) Interleukin 1 induces endothelial cell synthesis of plasminogen activator inhibitor. J Exp Med 163: 1595–1604
Nawroth PP, Stern DM (1986) Modulation of endothelial cell hemostatic properties by tumor necrosis factor. J Exp Med 163: 740–745
Neill RJ, Gemski P, Formal SB, Newland JW (1988) Deletion of the Shiga toxin gene in a chlorate-resistant derivative of *Shigella dysenteriae* type 1 that retains virulence. J Infect Dis 158: 737–741
Neisser M, Shiga K (1903) Ueber freie Receptoren von Typhus- und Dysenterie-Bazillen und ueber das Dysenterie Toxin. Dtsch Med Wochenschr 29: 61–62
Newland JW, Neill RJ (1988) DNA probes for Shiga-like toxins I and II and for toxin-converting bacteriophage. J Clin Microbiol 26: 1292–1297
Newland JW, Strockbine NA, Miller SF, O'Brien AD, Holmes RK (1985) Cloning of Shiga-like toxin structural genes from a toxin converting phage of *Escherichia coli*. Science 230: 179–181
Newland JW, Strockbine NA, Neill RJ (1987) Cloning of genes for production of *Escherichia coli* Shiga-like toxin type II. Infect Immun 55: 2675–2680
O'Brien AD, Holmes RK (1987) Shiga and Shiga-like toxins. Microbiol Rev 51: 206–220
O'Brien AD, LaVeck GD (1983) Purification and characterization of a *Shigella dysenteriae* 1-like toxin produced by *Escherichia coli*. Infect Immun 40: 675–683
O'Brien AD, LaVeck GD, Griffin DE, Thompson MR (1980) Characterization of *Shigella dysenteriae* 1 (Shiga) toxin purified by anti-Shiga toxin affinity chromatography. Infect Immun 30: 380–384
O'Brien AD, LaVeck GD, Thompson MR, Formal SB (1982) Production of *Shigella dysenteriae* type 1-like cytotoxin by *Escherichia coli*. J Infect Dis 146: 763–769
O'Brien AD, Lively TA, Chen ME, Rothman SW, Formal SB (1983) *Escherichia coli* O157: H7 strains associated with haemorrhagic colitis in the United States produce *Shigella dysenteriae* 1 (Shiga) like cytotoxin. Lancet i: 702
O'Brien AD, Newland JW, Miller SF, Holmes RK, Smith HW, Formal SB (1984) Shiga-like toxin-converting phages from *Escherichia coli* strains that cause hemorrhagic colitis or infantile diarrhea. Science 226: 694–696
O'Brien AD, Marques LRM, Kerry CF, Newland JW, Holmes RK (1989) Shiga-like toxin converting phage of enterohemorrhagic *Escherichia coli* strain 933. Microb Pathog 6: 381–390
Obrig TG, Moran TP, Brown JE (1987) The mode of action of Shiga toxin on peptide elongation of eukaryotic protein synthesis. Biochem J 244: 287–294
Obrig TG, DelVecchio PJ, Brown JE, Moran TP, Rowland BM, Judge TK, Rothman SW (1988) Direct cytotoxic action of Shiga toxin on human vascular endothelial cells. Infect Immun 56: 2373–2378
Olsnes S, Eiklid K (1980) Isolation and characterization of *Shigella shigae* cytotoxin. J Biol Chem 225: 284–289
Olsnes S, Sandvig K (1988) How protein toxins enter and kill cells. In: Frankel AE (ed) Immunotoxins. Kluwer, Boston, pp 39–73
Olsnes S, Fernandez-Puentes C, Carrasco L, Vazquez D (1975) Ribosome inactivation by the toxic lectins abrin and ricin. Kinetics of the enzymatic activity of the toxin A-chains. Eur J Biochem 60: 281–288
Olsnes S, Reisbig R, Eiklid K (1981) Subunit structure of Shigella toxin. J Biol Chem 256: 8732–8738
Ostroff SM, Tarr PI, Neill MA, Lewis JH, Hargrett-Bean N, Kobayashi JM (1989) Toxin genotypes and plasmid profiles as determinants of systemic sequelae in *Escherichia coli* O157: H7 infections. J Infect Dis 160: 994–999
Pai CH, Kelly JK, Meyers GL (1986) Experimental infection of infant rabbits with verotoxin-producing *Escherichia coli*. Infect Immun 51: 16–23
Perera LP, Samuel JE, Holmes RK, O'Brien AD (1991) Identification of three amino acid residues in the B subunit of Shiga toxin and Shiga-like toxin II essential for holotoxin activity. J Bacteriol 173: 1151–1160
Prywes R, Livneh E, Ullrich A, Schlessinger J (1986) Mutations in the cytoplasmic domain of EGF receptor affect EGF binding and receptor internalization. EMBO J 5: 2179–2190
Reisbig R, Olsnes S, Eiklid K (1981) The cytotoxic activity of Shigella toxin. Evidence for catalytic inactivation of the 60 S ribosomal subunit. J Biol Chem 256: 8739–8744

Remis RS, MacDonald KL, Riley LW, Phur ND, Wells JG, David BR, Blake PA, Cohen ML (1984) Sporadic cases of hemorrhagic colitis associated with *Escherichia coli* O157: H7. Ann Intern Med 101: 624–626

Rietra PJGM, Willshaw GA, Smith HR, Field AM, Scotland SM, Rowe B (1989) Comparison of Vero cytotoxin-encoding phages from *Escherichia coli* of human and bovine origin. J Gen Microbiol 135: 2307–2318

Riley LW, Remis RS, Helgerson SD, McGee HB, Wells JG, Davis BR, Hebert RJ, Olcott ES, Johnson LM, Hargrett NT, Blake PA, Cohen ML (1983) Hemorrhagic colitis associated with a rare *Escherichia coli* serotype. N Engl J Med 308: 681–685

Rose PE, Armour JA, Williams CE, Hill FGH (1985) Verotoxin and neuraminidase induced platelet aggregating activity in plasma; their possible role in the pathogenesis of the haemolytic uraemic syndrome. J Clin Pathol 38: 438–441

Rothenberger S, Iacopetta BJ, Kühn LC (1987) Endocytosis of the transferrin receptor requires the cytoplasmic domain but not its phosphorylation site. Cell 49: 423–431

Royall JA, Berkow RL, Beckman JS, Cunningham MK, Matalon S, Freeman BA (1989) Tumor necrosis factor and interleukin 1α increase endothelial permeability. Am J Physiol 257: L399–L410

Ryd M, Alfredsson H, Blomberg L, Andersson A, Lindberg AA (1989) Purification of Shiga toxin by α-D-galactose-(1-4)-β-D-galactose-(1-4)-β-D-glucose-(1)-receptor ligand-based chromatography. FEBS Lett 258: 320–322

Samuel JE, Perera LP, Ward S, O'Brien AD, Ginsburg V, Krivan HC (1990) Comparison of the glycolipid receptor specificities of Shiga-like toxin type II and Shiga-like toxin type II variants. Infect Immun 58: 611–618

Sandvig K, Brown JE (1987) Ionic requirements for entry of Shiga toxin from *Shigella dysenteriae* 1 into cells. Infect Immun 55: 298–303

Sandvig K, Olsnes S (1982) Entry of the toxic proteins abrin, modeccin, ricin, and diphtheria toxin into cells. I. Requirement for calcium. J Biol Chem 257: 7495–7503

Sandvig K, Tønnessen TI, Olsnes S (1986) Ability of inhibitors of glycosylation and protein synthesis to sensitize cells to abrin, ricin, Shigella toxin, and Pseudomonas toxin. Cancer Res 46: 6418–6422

Sandvig K, Olsnes S, Petersen OW, van Deurs B (1987) Acidification of the cytosol inhibits endocytosis from coated pits. J Cell Biol 105: 679–689

Sandvig K, Olsnes S, Petersen OW, van Deurs B (1988) Inhibition of endocytosis from coated pits by acidification of the cytosol. J Cell Biochem 36: 73–81

Sandvig K, Olsnes S, Brown JE, Petersen OW, van Deurs B (1989a) Endocytosis from coated pits of Shiga toxin: a glycolipid-binding protein from *Shigella dysenteriae* 1. J Cell Biol 108: 1331–1343

Sandvig K, Olsnes S, Petersen OW, van Deurs B (1989b) Control of coated-pit function by cytoplasmic pH. In: Tartakoff AM (ed) Methods in cell biology, vol 32. Vesicular transport, part B. Academic, New York, pp 365-382

Sandvig K, Prydz K, Ryd M, van Deurs B (1991a) Endocytosis and intracellular transport of the glycolipid-binding ligand Shiga toxin in polarized MDCK cells. J Cell Biol 113: 553–562

Sandvig K, Prydz K, van Deurs B (1991b) Endocytic uptake of ricin and Shiga toxin. Springer; Berlin Heidelberg New York

Sansonetti PJ (1991) Genetic and molecular basis of epithelial cell invasion by *Shigella* species. Rev Infect Dis 13 [Suppl 4]: S285–S292

Saxena SK, O'Brien AD, Ackerman EJ (1989) Shiga toxin, Shiga-like toxin II variant, and ricin are all single-site RNA N-glycosidases of 28S RNA when microinjected into *Xenopus* oocytes. J Biol Chem 264: 596–601

Schleimer RP, Rutledge BK (1986) Cultured human vascular endothelial cells acquire adhesiveness for neutrophils after stimulation with interleukin 1, endotoxin, and tumor-promoting phorbol diesters. J Immunol 136: 649–654

Schlossman D, Withers D, Welsh P, Alexander A, Robertus J Frankel A (1989) Role of glutamic acid 177 of the ricin toxin A chain in enzymatic inactivation of ribosomes. Mol Cell Biol 9: 5012–5021

Schmitt CK, McKee ML, O'Brien AD (1991) Two copies of Shiga-like toxin II-related genes common in enterohemorrhagic *Escherichia coli* strains are responsible for the antigenic heterogeneity of the O157: H strain E32511. Infect Immun 59: 1065–1073

Scotland SM, Smith HR, Willshaw GA, Rowe B (1983) Vero cytotoxin production in a strain of *Escherichia coli* determined by genes carried on bacteriophage. Lancet ii: 216

Seidah NG, Donohue-Rolfe A, Lazure C, AuClair F, Keusch GT, Chretien M (1986) Complete amino acid sequence of *Shigella* toxin B-chain. J Biol Chem 261: 13928–13931

Sekizaki T, Harayama S, Timmis KN (1987) Genetic manipulation *in vivo* of *stx*, a determinant essential for high level production of Shiga toxin by *Shigella dysenteriae* serotype 1: localization near *pyrF* and generation of stx^- transposon mutants. Infect Immun 55: 2208–2214

Shiga K (1898) Ueber den Dysenterie-bacillus (*Bacillus dysenteriae*). Zentralbl Bakteriol Orig 24: 913–918

Shite S, Seguchi T, Mizoguchi H, Ono M, Kuwano M (1990) Differential effects of brefeldin A on sialylation of N- and O-linked oligosaccharides in low density lipoprotein receptor and epidermal growth factor receptor. J Biol Chem 265: 17385–17388

Smith HW, Linggood MA (1971) Transmissible nature of enterotoxin production in a human enteropathogenic strain of *Escherichia coli*. J Med Microbiol 4: 301–305

Smith HW, Green P, Parsell Z (1983) Vero cell toxins in *Escherichia coli* and related bacteria: transfer by phage and conjugation and toxic action in laboratory animals, chickens, and pigs. J Gen Microbiol 129: 3121–3137

Strockbine NA, Marques LRM, Newland JW, Smith HW, Holmes RK, O'Brien AD (1986) Two toxin-converting phages from *Escherichia coli* O157: H7 strain 933 encode antigenically distinct toxins with similar biologic activities. Infect Immun 53: 135–140

Strockbine NA, Jackson MP, Sung LM, Holmes RK, O'Brien AD (1988) Cloning and sequencing of the genes for Shiga toxin from *Shigella dysenteriae* type 1. J Bacteriol 170: 1116–1122

Sung LM, Jackson MP, O'Brien AD, Holmes RK (1990) Analysis of transcription of the Shiga-like toxin type II and Shiga-like toxin type II variant operons of *Escherichia coli*. J Bacteriol 172: 6386–6395

Takami N, Oda K, Fujiwara T, Ikehara Y (1990) Intracellular accumulation and oligosaccharide processing of alkaline phosphatase under disassembly of the Golgi complex caused by brefeldin A. Eur J Biochem 194: 805–810

Takao T, Tanabe T, Hong Y-M, Shimonishi Y, Kurazono H, Yutsudo T, Sasakawa C, Yoshikawa M, Takeda Y (1988) Identity of molecular structure of Shiga-like toxin I (VT1) from *Escherichia coli* O157: H7 with that of Shiga toxin. Microb Pathog 5: 357–369

Tesh VL, O'Brien AD (1991) The pathogenic mechanisms of Shiga toxin and the Shiga-like toxins. Mol Microbiol 5: 1817–1822

Tesh VL, Samuel JE, Perera LP, Sharefkin JB, O'Brien AD (1991) Evaluation of the role of Shiga and Shiga-like toxins in mediating direct damage of human vascular endothelial cells. J Infect Dis 164: 344–352

Thompson MR, Steinberg MS, Gemski P, Formal SB, Doctor BP (1976) Inhibition of *in vitro* protein synthesis by *Shigella dysenteriae* 1 toxin. Biochem Biophys Res Commun 71: 783–788

Timmis KN, Clayton CL, Sekizaki T (1985) Localization of Shiga toxin gene in the region of *Shigella dysenteriae* 1 chromosome specifying virulence functions. FEMS Lett 30: 301–305

Tran D, Carpentier JL, Sawano F, Gorden P, Orci L (1987) Ligands internalized through coated or noncoated invaginations follow a common intracellular pathway. Proc Natl Acad Sci USA 84: 7957–7961

van Deurs B, Petersen OW, Olsnes S, Sandvig K (1987) Delivery of internalized ricin from endosomes to cisternal Golgi elements is a discontinuous, temperature-sensitive process. Exp Cell Res 171: 137–152

van Deurs B, Petersen OW, Olsnes S, Sandvig K (1989) The ways of endocytosis. Int Rev Cytol 117: 131–177

van Heyningen WE, Gladstone GP (1953a) The neurotoxin of *Shigella shigae*. 1. Production, purification and properties of the toxin. Br J Exp Pathol 34: 202–216

van Heyningen WE, Gladstone GP (1953b) The neurotoxin of *Shigella shigae*. 3. The effect of iron on production of the toxin. Br J Exp Pathol 34: 221–229

van Hinsbergh VWM, Kooistra T, Vanderberg EA, Princer HMG, Fiers W, Emeis JJL (1988) Tumor necrosis factor increases the production of plasminogen activator inhibitor in human endothelial cells *in vitro* and it rats *in vivo*. Blood 72: 1467–1473

Vicari G, Olitzki AL, Olitzki Z (1960) The action of the thermolabile toxin of *Shigella dysenteriae* on cells cultivated *in vitro*. Br J Exp Pathol 41: 179–189

von Wulfen H, Russmann H, Karch H, Meyer T, Bitzan M, Kohrt TC, Aleksic S (1989) Verocytotoxin-producing *Escherichia coli* O2: H5 isolated from patients with ulcerative colitis. Lancet i: 1449–1450

Waddell T, Head S, Petric M, Cohen A, Lingwood C (1988) Globotriosyl ceramide is specifically recognized by the *Escherichia coli* verocytotoxin 2. Biochem Biophys Res Commun 152: 674–679

Waddell T, Cohen A, Lingwood CA (1990) Induction of verotoxin sensitivity in receptor-deficient cell lines using the receptor glycolipid globotriaosyl-ceramide. Proc Natl Acad Sci (USA) 87: 7898–7901

Wadolkowski EA, Burris JA, O'Brien AD (1990a) Mouse model for colonization and disease caused by enterohemorrhagic *Escherichia coli*. Infect Immun 58: 2438–2445

Wadolkowski EA, Sung LM, Burris JA, Samuel JE, O'Brien AD (1990b) Acute renal tubular necrosis and death of mice orally infected with *Escherichia coli* strains that produce Shiga-like toxin type II. Infect Immun 58: 3959–3965

Weinstein DL, Holmes RK, O'Brien AD (1988a) Effect of iron and temperature on Shiga-like toxin I production by *Escherichia coli*. Infect Immun 56: 106–111

Weinstein DL, Jackson MP, Samuel JE, Holmes RK, O'Brien AD (1988b) Cloning and sequencing of a Shiga-like toxin II variant from an *Escherichia coli* strain responsible for edema disease of swine. J Bacteriol 170: 4223–4230

Weinstein DL, Jackson MP, Perera LP, Holmes RK, O'Brien AD (1989) In vivo formation of hybrid toxins comprised of Shiga toxin and the Shiga-like toxins and the role of the B subunit in localization and cytotoxic activity. Infect Immun 57: 3743–3750

Wells JG, Davis BR, Wachsmuth IK, Riley LW, Remis RS, Sokolow R, Morris GK (1983) Laboratory investigation of hemorrhagic colitis outbreaks associated with a rare *Escherichia coli* serotype. J Clin Microbiol 18: 512–520

Willshaw GA, Smith HR, Scotland SM, Field AM, Rowe B (1985) Cloning of genes determining the production of Vero cytotoxin by *Escherichia coli*. J Gen Microbiol 131: 3047–3053

Willshaw GA, Smith HR, Scotland SM, Field AM, Rowe B (1987) Heterogeneity of *Escherichia coli* phages encoding Vero cytotoxins: comparison of cloned sequences determining VT1 and VT2 and development of specific gene probes. J Gen Microbiol 133: 1309–1317

Yutsudo T, Hondo T, Miwatani T, Takeda Y (1986) Characterization of purified Shiga toxin from *Shigella dysenteriae* I. Microbiol Immunol 30: 1115–1127

Environmental Regulation of *Shigella* Virulence*

A. T. MAURELLI[1], A. E. HROMOCKYJ[1,3] and M. L. BERNARDINI[2,4]

1 Introduction

Bacterial pathogens frequently find themselves exposed to a variety of diverse, frequently hostile conditions during their infectious cycle. The journey from the external environment to the host, where the bacteria may ultimately cause disease, is an adventurous one and involves exposure of the organism to a multitude of growth conditions. Successful pathogens have evolved the means for survival within the varied growth conditions encountered both inside and outside their respective hosts. The ability to synthesize proteins which

---

* The opinions or assertions of A.T.M. and A.E.H. contained herein are the private ones of the authors and are not to be construed as official or reflecting the views of the Department of Defense or the Uniformed Services University of the Health Sciences

[1] Department of Microbiology, Uniformed Services University of the Health Sciences, 4301 Jones Bridge Road, Bethesda, MD 20814-4799, USA
[2] Unité de Pathogénie Microbienne Moléculaire, Institut Pasteur, 28 rue du Docteur Roux, 75724 Paris CEDEX 15, France
[3] Present Address: Department of Microbiology and Immunology, Stanford University School of Medicine, Stanford, CA 94305, USA
[4] Present Address: Dip. di Biologia Cellulare e dello Sviluppo, Sezione Scienze Microbiologiche, Via degli Apuli, 1, 00195 Roma, Italia

Current Topics in Microbiology and Immunology, Vol. 180

facilitate survival under different growth conditions is part of an organism's adaptive response. The adaptive response also involves mechanisms for sensing environmental changes and, in turn, regulating gene expression. Many examples exist of nonpathogenic bacteria which regulate genes in response to changes in their surroundings. Nutrient limitation, a shift in growth from one carbon source to another, a change from aerobic to anaerobic growth conditions, and changes in osmolarity and temperature are all signals to which bacteria respond by altering gene expression (reviewed by GOTTESMAN 1984; STOCK et al. 1989).

An increasingly common theme being found among bacterial pathogens is the coordinate regulation of virulence gene expression in response to environmental stimuli (reviewed by JF MILLER et al. 1989; DIRITA and MEKALANOS 1989). Many bacterial pathogens regulate virulence gene expression in response to signals such as iron limitation, oxygen limitation, pH, temperature, osmolarity, and calcium levels. Such environmental cues may act to signal to the organism that it has found the proper host niche. Bacterial pathogens have also evolved systems which place expression of their virulence genes within a global regulatory control circuit which is responsive to external stimuli. This strategy solves two fundamental metabolic problems for the bacterium. First, the system allows the organism to conserve energy by repressing the expression of virulence genes when these gene products are not needed, e.g., outside the host. Second, coordinate regulation permits the simultaneous expression of unlinked genes in response to an environmental stimulus. Thus bacterial pathogens are highly specialized organisms capable of monitoring and adapting to changes in their surroundings. The economical expression of a complex network of genes permits their survival within or outside the host.

We have been studying the regulation of virulence gene expression in *Shigella flexneri* 2a as a model for understanding regulation of bacterial pathogenesis. Studies which employ virulence assays along with classical genetic techniques and recombinant DNA technology have revealed that the virulence of *Shigella* species is a multigenic phenomenon. Virulence loci have been identified which are encoded both chromosomally and on a large 180- to 220-kilobase (kb) virulence plasmid. Thus, expression of virulence in *Shigella* presents an ideal example of the need for a global regulatory system to exert control over many unlinked plasmid and chromosomal virulence genes and to regulate them coordinately. The principal environmental stimulus for virulence gene regulation in *Shigella* is temperature, and, indeed, temperature regulation of virulence gene expression is a hallmark of *Shigella* pathogenesis. However, recent reports have presented evidence which demonstrate alterations in the expression of *Shigella* virulence in response to other changes in growth conditions. Studies of the molecular basis of these observed phenotypic changes have led to the formulation of an expanding network of regulatory genes which act to modulate the expression of *Shigella* virulence genes. The purpose of this chapter is to review what we now know of several environmental stimuli which influence the expression of virulence genes in *Shigella*.

2 Temperature Regulation of Virulence Gene Expression

2.1 Definition and Extent of Regulon

Temperature is an obvious environmental cue which could be exploited by human pathogens to trigger the regulation of virulence genes. The temperature within the human host is generally 37 °C. Thus, a shift from 25 °C to 37 °C could indicate to the bacteria that virulence genes should now be turned on. *Shigella* uses this change in temperature as a signal to regulate the expression of its virulent phenotype (Table 1; reviewed by MAURELLI 1989a). Although *Shigella* is not unique in regulating virulence gene expression in response to temperature, the molecular mechanism of temperature-regulated virulence gene expression in *Shigella* may serve as a model system of gene regulation in response to temperature (MAURELLI 1989b).

Temperature was first described as an environmental signal which effects the expression of *Shigella* virulence when it was demonstrated that wild-type strains are fully virulent and invasive when cultured at 37 °C, but become phenotypically avirulent and noninvasive when grown at 30 °C (MAURELLI et al. 1984a). Virulent strains of *S. flexneri*, *S. sonnei*, and *S. dysenteriae* which have been grown at 30 °C are neither able to invade epithelial cells nor to evoke keratoconjunctivitis in guinea pigs. Yet by shifting the growth temperature to 37 °C, virulence is fully restored. Re-expression of virulence requires several generations of growth at 37 °C and is dependent on de novo protein synthesis.

Table 1. Temperature regulation of virulence-associated phenotypes of *Shigella*

Virulence-associated phenotype	Regulation by growth temperature
Maintenance of virulence plasmid	No
Invasion of mammalian cells in culture	Yes
Invasion and intracellular multiplication (plaque assay)	Yes
Invasion and keratoconjunctivitis production in guinea pig (Sereny test)	Yes
Contact hemolytic activity	Yes
Pigmentation on Congo red agar	Yes
Aerobactin production	No
Shiga toxin production (*S. dysenteriae* 1)	Yes
Synthesis of *ipa* gene products (*i*nvasion *p*lasmid *a*ntigen)	
IpaA (78.0 kDa)	Yes
IpaB (62.1 kDa)	Yes
IpaC (38.7 kDa)	Yes
IpaD (38.0 kDa)	Yes
Synthesis of *mxi* gene products (*m*embrane expression of *i*nvasion plasmid antigens)	Yes
Expression of the positive regulators *virF* and *virB*	Yes

The temperature regulation of *Shigella* virulence genes is not a heat-shock response in the generally accepted sense of this well-defined phenomenon. The heat-shock response is characterized by an immediate but transient expression of a subset of genes in response to elevated temperature. Exposure to other forms of cell-damaging agents such as viruses, oxygen radicals, ethanol, etc. also provoke expression of the heat-shock response (NEIDHARDT and VANBOGELEN 1987). In contrast, the modulation of virulence gene expression in *Shigella* is mediated by temperature in a fashion which involves derepression of a set of genes and steady-state expression of these genes at the permissive temperature, i.e., 37 °C. Thus, whereas the heat-shock response is a cellular reaction to potentially lethal environmental insult, temperature regulation in *Shigella* is a form of "on-off" switch for expression of the virulence regulon only under conditions where such gene expression is necessary for competing effectively in the mammalian host.

Operon fusion technology has proven to be an effective tool for analyzing the genetics of temperature regulation and for determining the level at which regulation by temperature occurs. Random insertions of an operon fusion phage carrying a promoterless *lacZ* gene were generated in a virulent strain of *S. flexneri* serotype 2a (MAURELLI and CURTISS 1984). *lacZ* codes for the enzyme *β*-galactosidase, and formation of positive operon fusions results in *lacZ* being placed under the transcriptional control of the target gene promoter. *β*-galactosidase expression can then serve as an easy way to monitor and quantitate the transcriptional activity of a target gene promoter. By using this strategy, a plasmid-encoded virulence gene (*vir*) was fused to *lacZ*, and the loss of epithelial cell invasion was shown to be directly linked to temperature-regulated expression of *β*-galactosidase from the *vir*::*lacZ* mutant. This indicates that virulence gene regulation by temperature is at the transcriptional level.

The generation of random operon fusions in *S. flexneri* also allowed for the identification of novel virulence genes by employing a selection for temperature-regulated expression of *lacZ* (i.e., Lac$^+$ at 37 °C and Lac$^-$ at 30 °C) and screening for simultaneous loss of virulence. The *mxi* (*m*embrane e*x*pression of *i*nvasion plasmid antigens) loci on the virulence plasmid were identified by using this strategy (HROMOCKYJ and MAURELLI 1989a; ANDREWS et al. 1991). These mutants are noninvasive, altered in surface expression of the Ipa virulence gene products, and fail to excrete these proteins into the medium as the wild-type parent does. Thus it has been possible to exploit the temperature-regulated nature of *Shigella* virulence to identify new genes essential to *Shigella* pathogenesis.

2.2 Negative Regulator—*virR*

The power of operon fusion technology also permits the design of strategies for the analysis of gene regulation. In this case, the *vir*::*lacZ* operon fusion described above was used to select for mutants which constitutively express the virulence gene independent of temperature. Random Tn*10* mutagenesis of the *vir*::*lacZ* strain led to the identification of *virR*, a chromosomally encoded locus required for temperature regulation (MAURELLI and SANSONETTI 1988). The *virR*::Tn*10* mutants which were identified constitutively express β-galactosidase from the *vir*::*lacZ* reporter gene fusion at 30 °C. Transduction of the *virR*::Tn*10* mutation into a virulent strain of *S. flexneri* results in transductants which are deregulated for invasion of HeLa cells and are thereby able to invade even when grown at 30 °C. The *virR*::Tn*10* transductants also exhibit deregulated phenotypes in the plaque assay, Sereny test, and contact hemolysis assay. In addition, expression of the invasion plasmid antigens (IpaBCDA) is deregulated at 30 °C. Mapping of the *virR*::Tn*10* insertion placed the *virR* locus between *galU* and the *trp* operon at 27 min on the *S. flexneri* chromosome. The strategy for cloning the *virR* locus of *Shigella* exploited the close genetic linkage of *virR* and *galU*. A cosmid clone from a *S. flexneri* 2a library was selected for complementation of a *galU* mutation and was shown to restore temperature-regulated β-galactosidase expression in a *vir*::*lacZ* mutant harboring a *virR* deletion mutation. *virR* complementing activity maps to a 1.8-kb *EcoRI-AccI* fragment within this clone.

An overall DNA homology of approximately 90% exists between *Shigella* and *Escherichia coli* (BRENNER et al. 1972), and comparison of structural gene DNA sequence (BRAUN and COLE 1982; COSSART et al. 1986) and electrophoretic analysis of enzyme polymorphisms (OCHMAN et al. 1983) demonstrate the close genetic relationship between these two genera. In addition, enteroinvasive *E. coli* (EIEC) are also temperature regulated for virulence. Thus, laboratory strains of *E. coli* K-12 were tested and shown to carry a gene able to regulate the expression of a temperature-regulated virulence gene in *S. flexneri*. Therefore, it was concluded that a *virR* homologue exists in *E. coli* K-12 (HROMOCKYJ and MAURELLI 1989b).

In *E. coli* K-12, several regulatory loci (*drdX*, *bglY*, *osmZ*, and *pilG*) also map to 27 min, and each one has subsequently been identified as an allele of *hns*. a gene which encodes the histone-like protein (HLP) H-NS (GÖRANSSON et al. 1990; MAY et al. 1990; HULTON et al. 1990; KAWULA and ORNDORFF 1991; PON et al. 1988). Nucleotide sequence analysis of a portion of the *S. flexneri virR* clone confirmed the presence of a 412 bp ORF which exhibits over 99% sequence identity with the *E. coli hns* coding sequence. Insertion mutagenesis of the cloned *S. flexneri* ORF at a unique *HpaI* restriction site abolishes *virR*-complementing activity expressed by the VirR$^+$ clone. The cloned *pilG* locus of *E. coli* K-12 also complements a *virR* mutant of *S. flexneri*. These experiments present direct genetic evidence that *virR* of *S. flexneri* is an allele of *hns*. The demonstration that *virR* is allelic to *hns* raises the question as to how a protein such as VirR/H-NS can mediate the ability of shigellae to respond to a change in temperature and translate this change into a coordinately regulated virulence gene response. Models describing the possible molecular mechanisms of VirR/H-NS activity are discussed below.

2.3 Models of *virR/hns* Activity

H-NS and several other prokaryotic DNA-binding proteins have been characterized as HLPs on the basis of properties they share with eukaryotic histones (DRLICA and ROUVIERE-YANIV 1987). However, unlike eukaryotic histones, no evidence exists to indicate that H-NS or any of the bacterial HLPs forms analogous nucleosome structures. Biochemical analysis of H-NS indicates that it is an abundant (approximately 20000 copies per cell), low molecular weight (15 kDa), heat-stable protein which is one of the major protein components extracted along with *E. coli* chromatin (FALCONI et al. 1988). H-NS also exhibits a high binding affinity for double-stranded DNA, an ability to compact DNA in vitro as well as to increase the thermal stability of DNA (GUALERZI et al. 1986; FRIEDRICH et al. 1988). Comparison of the *hns* alleles from *E. coli* (GÖRANSSON et al. 1990), *Salmonella typhimurium* (HULTON et al. 1990; MARSH and HILLYARD 1990), *Serratia marcescens, Proteus vulgaris* (LATEANA et al. 1989), and *S. flexneri* (A.E. HROMOCKYJ and A. T. MAURELLI, manuscript submitted) shows that a high degree of nucleotide and predicted amino acid sequence homology exists among these genes. Therefore, *virR/hns* is a highly conserved gene which plays a central role in the regulation of various unlinked genes and operons within the enterobacteriaceae.

Two models have been proposed to explain the mechanism by which VirR/H-NS functions to regulate virulence gene expression in response to changes in temperature. One of these models invokes environmentally induced alterations in DNA supercoiling as a general mechanism of bacterial gene regulation in response to several different external stimuli (reviewed by HIGGINS et al. 1990). VirR/H-NS is believed to mediate alterations in DNA supercoiling by an as yet undetermined mechanism and, in turn, to repress gene transcription. Several studies have reported that virulence gene expression in *Salmonella typhimurium* (GALAN and CURTISS 1990) and *Shigella flexneri* (DORMAN et al. 1990) is modified when the bacteria are subjected to conditions which alter DNA supercoiling. All of the evidence presented in favour of this model is, however, indirect since patterns of DNA supercoiling were monitored as variations in reporter plasmid supercoiling as opposed to changes in DNA topology that occur adjacent to the reporter genes studied. Yet even if one accepts the assumption that reporter plasmid supercoiling reflects an overall change in bacterial DNA supercoiling, there are certain inconsistencies in the DNA supercoiling model. For example, the effects of growth temperature on plasmid DNA supercoiling in *E. coli* are opposite to the effects observed in *S. flexneri.* Also, identical *virR*::Tn*10* mutations in *E. coli* and *S. flexneri* exhibit opposite effects on reporter plasmid supercoiling in the two strains (DORMAN et al. 1990). Yet the *E. coli virR* homologue (presumably *hns*) regulates virulence gene expression in *S. flexneri* in a manner indistinguishable from the *S. flexneri virR* gene (HROMOCKYJ and MAURELLI 1989b). In addition, temperature-regulated expression of *pap* pilin in uropathogenic *E. coli*, which requires the activity of the *hns* allele *drdX*, is identical to the pattern of virulence gene regulation in shigellae (GÖRANSSON and UHLIN 1984). Therefore, uropathogenic *E. coli* and *Shigella* spp. regulate virulence gene expression in response to temperature change identically, despite the fact that both temperature and a mutation in *virR* each have different effects on DNA supercoiling in these two organisms. Thus, these experimental discrepancies need to be reconciled before DNA supercoiling can be established as a unifying model for environmentally regulated gene expression.

An alternative explanation for *virR*-mediated virulence gene regulation is that VirR may bind specific DNA sequences, inhibit RNA polymerase activity, and thereby repress transcription. The ability of the bacteria to detect temperature change and coordinate a regulatory response could be achieved by the temperature regulation of *virR* expression. In this model, *virR* expression at 30 °C would result in synthesis of the repressor which could bind regulatory sequences and repress virulence gene transcription. Conversely, the lack of *virR* expression at 37 °C would result in the absence of the regulatory protein and the subsequent derepression of the virulence phenotype. In vitro evidence in support of this model is the fact that H-NS strongly inhibits DNA transcription, specifically transcription initiation (GUALERZI et al. 1986; SPASSKY et al. 1984). Further, the identification of H-NS as one of 14 cold-shock proteins of *E. coli* (JONES et al. 1987; VANBOGELEN et al. 1990; VANBOGELEN and NEIDHARDT 1990) supports the notion that VirR synthesis is temperature regulated. H-NS synthesis is detected in bacteria grown at 37 °C, and, following a downshift in bacterial growth temperature from 37 °C to 10 °C, the relative rate of H-NS synthesis increases. It is conceivable that a reduction in temperature from 37 °C to 30 °C might also account for a similar, albeit not as dramatic, increase in VirR synthesis. Studies with the *E. coli hns* allele *drdX* demonstrate that the amount of *drdX*-specific mRNA is unaltered when purified from strains grown at either 25 °C or 37 °C (GÖRANSSON et al. 1990). This observation would indicate that *virR* expression may be unaffected by temperature change, at least at the transcriptional level. Since altered ribosome function can induce the synthesis of cold-shock proteins including H-NS (VANBOGELEN and NEIDHARDT 1990), induction of VirR/H-NS synthesis after cold shock may be the result of an altered translational capacity of the bacteria, and the ribosome may be the sensor which detects changes in temperature.

Another possible explanation for the absence of VirR activity at 37 °C is that the protein itself is labile at this temperature and impaired in its activity as a repressor. Although the description of VirR/H-NS as a heat-stable protein which binds nucleic acids in vitro at 37 °C might argue against this possibility (GUALERZI et al. 1986), it is also possible that binding behaves differently in vivo. Alternatively, if VirR/H-NS forms oligomers, as suggested by protein cross-linking studies (SPASSKY and BUC 1977; GUALERZI et al. 1986; FALCONI et al. 1988). then it may be the oligomerization of VirR/H-NS which is temperature labile at 37 °C. Moreover, specificity of in vivo binding to target gene regulatory sequences may depend on stable VirR oligomer formation at 30 °C. Instability of oligomers at 37 °C would then allow for nonspecific VirR monomer binding of DNA and relief of transcription repression. On the other hand, oligomerization of VirR might be concentration dependent and favoured by an increased intracellular pool of VirR monomer. This model is suggested by H-NS protein cross-linking experiments in which H-NS dimers, trimers, and tetramers are formed as the concentration of H-NS increases (SPASSKY and BUC 1977; GUALERZI et al. 1986; FALCONI et al. 1988). As mentioned above, the formation of oligomers could facilitate DNA sequence-specific binding of VirR and thus inhibit virulence gene transcription. At 37 °C, lower levels of translation would result in lower levels of VirR monomer synthesis and thus preclude oligomer formation. Although monomers could nonspecifically bind to DNA under these conditions, transcription would be unaffected.

These models are speculative, and clearly further study is necessary to determine which is correct. The increasing availability of cloned temperature-regulated genes and promoters from *Shigella* spp. (HROMOCKYJ and MAURELLI 1989a: ANDREWS et al. 1991; TOBE et al. 1991) provides a potential source of specific VirR DNA-binding sites that could be used to test the models proposed above.

Characterization of the *virR* locus surprisingly revealed that partial complementation of a *ΔvirR* mutation could be attained through multicopy expression of the transfer RNA for tyrosine ($tRNA1^{tyr}$; A.E. HROMOCKYJ and A.T. MAURELLI, manuscript submitted), The VirR partial phenotype ($VirR^{P}$) is characterized as an inability to repress *β*-galactosidase expression from the *vir::lacZ* fusion in the ΔvirR reporter strain to the same levels as exhibited by the *virR* clones. Nucleotide sequence analysis of the $VirR^{P}$ clone revealed a high degree of nucleotide sequence homology with the *tyrT* locus of *E. coli* K-12 (GOODMAN et al. 1968; SEKIYA et al. 1976). Deletion of either a part of or the entire cloned $tRNA1^{tyr}$ sequence results in loss of the $VirR^{P}$ phenotype. In addition, an identical $VirR^{P}$ phenotype is observed when a clone of the $tRNA1^{leu}$ operon is used to complement a *ΔvirR* mutation. Therefore, the phenotype appears to be the result of a nonspecific effect achieved by overexpression of at least two different tRNA species.

The finding that a component of the bacterial translational machinery is able to regulate gene expression at the transcriptional level is not unprecedented. Transcription attenuation and the stringent response are two examples where bacterial transcription is modulated by the availability of amino-acylated or charged tRNA molecules to incorporate amino acids into a newly synthesized polypeptide. However, neither transcriptional attenuation nor the stringent response serves to explain the $VirR^{P}$ phenotypes expressed by the tRNA clones. It is possible that overproduction of the $tRNA1^{tyr/leu}$ has an effect on the translational capacity of the bacteria transformed with the tRNA clones, thereby inducing a cold-shock response. Thus, the $VirR^{P}$ phenotype may further substantiate the model that bacterial sensing of temperature change is at the level of the translational machinery.

2.4 Positive Regulators—*virF* and *virB*

virF was originally identified as a 1.0-kb region of the *S. flexneri* virulence plasmid which is essential but not sufficient to restore a Pcr^+ (pigmentation on Congo red medium) phenotype in certain Pcr^- mutants of *Shigella* (SAKAI et al. 1986a). The *virF* gene was localized to the *Sal*I F restriction fragment of the virulence plasmid. Certain *Sal*I F::Tn5 mutants are noninvasive for HeLa cells and are Sereny test negative, which indicates that *virF* is associated with virulence. *virF* mutants exhibit reduced levels of VirG and IpaB, C, and D expression and wild-type levels of each of these proteins are restored by transformation of the mutant strain with a cloned *Sal*I F fragment (SAKAI et al. 1988). In the presence of cloned *virF*, *virG* mRNA levels and *β*-galactosidase expression from a *virG*::*lacZ* operon fusion are increased as well. *virF* therefore acts to positively regulate expression of both *virG* and the *ipaBCDA* operon. Minicell analysis indicates that three proteins of 30, 27, and 21 kDa are synthesized from the cloned *virF* and correspond in size to the open reading frames predicted from the DNA sequence (SAKAI et al. 1986b). The 30-kDa VirF product was later shown to be a transcriptional activator of *virG* expression (SAKAI et al. 1988). Southern hybridization of virulence plasmid DNA from *S. sonnei*, *Shigella boydii*, *S. dysenteriae*, and EIEC strains with a *virF*-specific probe revealed that sequences homologous to *virF* could be detected in all of these bacterial species. These results, along with the identification of a gene in *S. sonnei* identical in function and DNA sequence to the *S. flexneri virF* gene (KATO et al. 1989), suggest that the mechanisms of virulence gene activation are conserved in *Shigella* and EIEC. Interestingly, the predicted amino acid sequence of VirF does not exhibit significant protein sequence homology with any other known transcriptional regulator. The molecular mechanism by which VirF acts as a positive regulator is still unknown. Further, it remains to be determined whether VirF can bind directly to virulence gene promoters or if it acts in concert with other, as yet unidentified, proteins to activate gene expression.

The *virB* locus was first identified by Tn5 transposon insertions within the *Sal*I B restriction fragment of the *S. flexneri* 2a virulence plasmid (SASAKAWA et al. 1986). Specific *Sal*I B::Tn5 mutants are unable to invade epithelial cells, evoke a positive Sereny test, or bind Congo red. A DNA fragment capable of restoring a wild-type virulence phenotype to the *Sal*I::Tn5 mutants was isolated and designated *virB* (ADLER et al. 1989). Deletion mapping and complementation analysis located *virB* just downstream of the last gene of the *ipa* operon (*ipaA*) but in a separate transcriptional unit. Minicell analysis of the *virB* clone revealed that a single protein of 33 kDa was synthesized. Nucleotide sequence analysis of the *virB* recombinant identified a single ORF which corresponds to a protein with a predicted molecular weight of 35.4 kDa. Detailed characterization of *virB*::Tn5 mutants showed that expression of the essential virulence peptides IpaB and IpaC is undetectable in these strains and expression of IpaD is reduced, whereas the expression of VirG is unaltered. The block in expression of the *ipa* genes in the *virB*::Tn5 mutants is due to reduced levels of *ipaB*, *C*, and *D* mRNA. When these *virB*⁻ mutant strains are transformed with a cloned *virB* gene, they regain the ability to express wild-type levels of IpaB, C, and D. This also correlates with the ability of the transformants to invade epithelial cells. Taken together, the data indicate that restoration of wild-type IpaB, C, and D expression in the *virB*⁻ mutant is due to the activation of *ipa* transcription by the *virB* clone. Activation by *virB* also appears to be the sole basis for the temperature-dependent nature of *ipa* gene expression. When *virB* expression is placed under the control of an exogenous promoter (P*tac*) and induced at 30 °C, *ipa* mRNA synthesis is induced as well (TOBE et al. 1991).

Independent studies have identified genes from *S. flexneri* serotype 5 (*ipaR*) and from *S. sonnei* (*invE*) which are analogous in function and identical in their predicted protein sequences to *virB* (BUYSSE et al. 1990; WATANABE et al. 1990). Results from the *ipaR* study demonstrated that *ipaA* expression is also positively regulated by *ipaR*. Analysis of the predicted amino acid sequences of *ipaR* and *invE* reveal a striking amino acid sequence homology between these proteins and ParB of bacteriophage P1 and SopB of the F plasmid. Both ParB and SopB are essential for plasmid maintenance and partitioning to daughter cells and are DNA-binding proteins (ABELES et al. 1985; MORI et al. 1986). Yet mutations in *virB/ipaR/invE* do not appear to alter virulence plasmid stability in *Shigella*. Further, the *sopB* gene of a mini-F cloning vector cannot complement a mutant *invE* in *S. sonnei* (WATANABE et al. 1990). The amino acid sequence homologies may reflect the motifs of a DNA-binding protein rather than a shared function. However, InvE shows no significant homology with known activator proteins of the LysR, OmpR, and AraC families, and a computer search for matches between IpaR and the DNA-binding domains of TdcR, Fis, NtrC, TnpR, Hin, and the LysR family of *E. coli* regulatory proteins showed no significant similarities either (WATANABE et al. 1990; BUYSSE et al. 1990). Nevertheless, it appears that *virB/ipaR/invE* serves as a transcriptional activator of certain plasmid-encoded virulence genes and may regulate these genes through a DNA-binding mechanism. The actual binding of VirB to specific operator-promoter regions of virulence genes awaits experimental demonstration.

The initial characterization of *virB* also revealed that the level of *virB* transcripts in a *virF*::Tn5 mutant are reduced and that introduction of a *virF* clone restores wild-type levels of *virB* mRNA (ADLER et al. 1989). Thus it was concluded that expression of *virB* is dependent on *virF*. On the basis of these observations, a dual transcriptional activator model for the regulation of virulence gene expression in *Shigella* was proposed (ADLER et al. 1989). In this model, VirF indirectly regulates the *ipaBCD* genes through the transcriptional activation of *virB* which, in turn, directly activates expression of the *ipaBCD* genes (Fig. 1). S1 nuclease protection assays have demonstrated that expression of both *virF* and *virB* is regulated by growth temperature (TOBE et al. 1991). Transcription of *virB* is tightly controlled by temperature, and 20-fold greater amounts of mRNA are seen at 37 °C versus 30 °C. Synthesis of *virF* mRNA is less dramatically affected by temperature (four fold greater at 37 °C versus 30 °C). This modest induction of *virF* synthesis at 37 °C would not appear to be

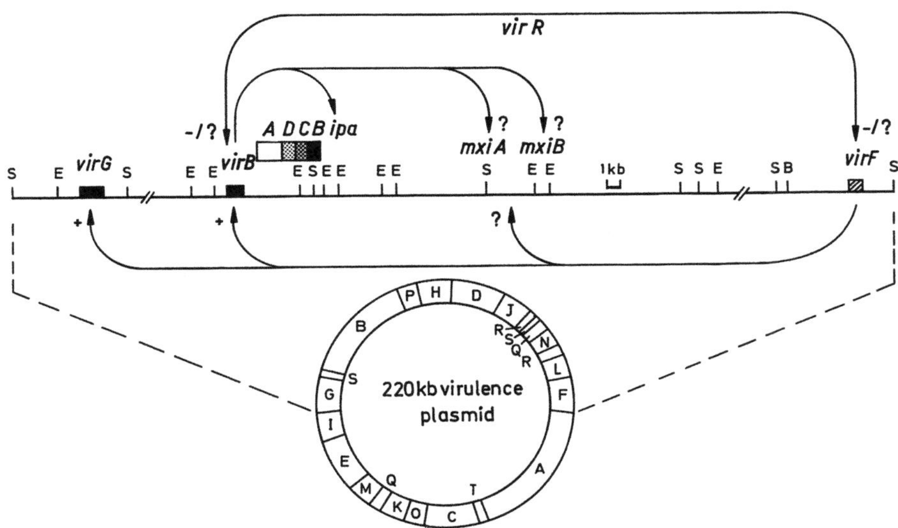

Fig. 1. Two-tiered regulatory loop for temperature-dependent regulation of *Shigella* virulence gene expression. *Top*, chromosomal *virR* locus; *middle*, portion of the *S. flexneri* virulence plasmid and selected virulence gene loci; *Bottom*, *SalI* restriction map of the virulence plasmid of *S. flexneri*. *Arrows*, genes regulated by *virR*, *virF*, and/or *virB*; *plus signs*, positive regulation; *minus signs*, negative regulation; *question marks*, the type of regulation is not known

sufficient to explain the striking induction of *virB* mRNA at this temperature. In fact, even when transcription of *virF* is artificially induced at 30 °C (via an external, inducible promoter), no concomitant induction of *virB* mRNA synthesis is seen (Tobe et al. 1991). Thus, although expression of *virB* is clearly dependent on *virF*, it is also very much dependent on temperature.

What further role might temperature be playing in the induction of *virB*? It is possible that VirF undergoes some sort of post-transcriptional modification at 37 °C which permits transcription of *virB*. Alternatively, an additional transcription factor encoded by a temperature-regulated gene may be required for the induction of *virB* mRNA synthesis. A third possibility involves the role of *virR*. The *virR* product may be directly involved in repressing the expression of *virF and virB* at 30 °C by binding to their operator-promoter regions. Both genetic and biochemical approaches should help to clarify further the nature of this increasingly complex regulatory circuit.

3 Osmotic Regulation of Virulence

Shigellae encounter conditions of high osmolarity in different compartments of the human host and, in addition, they have to adapt to changes in osmolarity in the passage from the extracellular state to the intracellular state. The osmolarity inside mammalian cells is about 300 mOsm, similar to that obtained when bacteria are grown in a high-osmolarity medium (e.g., tryptone–yeast extract–glucose or medium, A supplemented with 15% sucrose or 0.3 *M* NaCl). Therefore, it is possible to reproduce in vitro the same osmolarity conditions that shigellae encounter in vivo. In *E. coli*, several cellular phenotypes have been identified as being osmotically dependent. These include the porin composition of the outer membrane, and the expression and activity of transport systems for the major osmolytes (for a recent review see CSONKA 1989). In *Shigella*, two classes of genes respond to changes in osmolarity. The first class includes genes whose expression is required for physiological adaptation of the bacteria to osmolarity changes. Genes such as *proU*, *ompF*, *ompC*, and *kdp* belong to this class. The second class includes genes not directly involved in physiological adaptation but whose expression is optimized under high osmolarity conditions. It is conceivable that a mutation in a gene belonging to either of these two classes could affect the virulence of *Shigella*.

3.1 The Osmotic Environment and Porin Expression

The permeability of the outer membrane of *E. coli* is largely determined by channels formed by porin homotrimers. Bacteria respond to changes in the osmotic strength of the environment by changing the ratio of the two main porins, OmpF and OmpC, so that the total level of porin protein remains approximately constant (LUGTENBERG et al. 1976). OmpF, which makes slightly wider pores, predominates at low osmotic strength, and OmpC is favored at high osmotic strength (NIKAIDO and VAARA 1985; HANCOCK 1987). This regulation is mediated at the level of transcription by the *ompR* and *envZ* genes which constitute the *ompB* locus located at 75 min on the *E. coli* chromosome (COMEAU et al. 1985). OmpR and EnvZ together constitute a two-component system which senses changes in medium osmolarity and transmits the environmental stimulus (RONSON et al. 1987). OmpR-EnvZ belong to a family of bacterial signal transduction proteins which includes both sensor and regulatory proteins. *envZ* encodes the transmembrane sensory component which is a histidine protein kinase acting as the phosphodonor on the product of *ompR*, the transcriptional regulator (FORST et al. 1987, 1989; NORIOKA et al. 1986). Two-component regulatory systems such as this are found in both Gram-negative and Gram-positive bacteria and mediate regulatory responses to changes in sugar and nitrate availability, osmolarity, redox state, nitrogen, and phosphate status (STOCK et al. 1989). Analogous signal transduction proteins are also used by pathogenic bacteria such as *Bordetella pertussis* (ARICO et al. 1989), *S. typhimurium* (SI MILLER et al. 1989), *Agrobacterium tumifaciens* (LEROUX et al. 1987), and *Pseudomonas aeruginosa* (DERETIC et al. 1989) during invasion of their respective hosts.

The *Shigella* equivalent of the *E. coli ompB* locus also maps at 75 min on the *Shigella* chromosome (BERNARDINI et al. 1990). Regulation of *ompC-ompF* expression in *Shigella* is controlled by *ompR-envZ* in a manner similar to the *E. coli* system in that high osmolarity conditions favor the expression of OmpC in preference to OmpF (M.L. BERNARDINI, unpublished observations). The smaller pore of OmpC could reduce the diffusion of larger hydrophobic and negatively charged molecules. Consequently, under conditions of high osmolarity, which virulent shigellae can encounter inside cells or in a body compartment like the duodenum, OmpC can offer protection against negatively charged inhibitory molecules such as bile salts. In contrast, in an external environment which presents lower osmotic strength and lower concentrations of toxic substances and nutrients, the larger channel of OmpF may be more desirable.

Deletion mutants of *ompB* and insertion mutants of *envZ* do not express porins OmpC and OmpF. In *Shigella*, both of these types of mutants are severely affected in virulence. The introduction of either an *ompB* deletion (*ΔompB*) or an *envZ*::Tn*10* mutation into an *S. flexneri* serotype 5 strain causes a dramatic reduction in the ability to enter into HeLa cells; a limited capacity to survive intracellularly; and an attenuated, transient, and delayed response in the Sereny test (BERNARDINI et al. 1990). The expression of the Ipa proteins in both of the mutant strains, however, remains unaltered.

3.2 Role of the *ompB* Locus in Osmoregulation of Virulence

In addition to regulation of the chromosomal *ompF* and *ompC* porin genes, the *ompB* locus of *Shigella* also regulates virulence plasmid genes in response to changes in the osmolarity of the culture medium. The plasmid-encoded *vir*::*lacZ* operon fusion, which was used to demonstrate the temperature-dependent nature of *Shigella* virulence gene expression, also responds to changes in osmolarity. Gene expression from the *vir*::*lacZ* fusion increases three- to fourfold in high-osmolarity conditions. A similar range of variation is observed with *ompC* and *ompF* expression under equivalent conditions. Osmotic pressures that induce submaximal expression of the *vir*::*lacZ* fusion are found in the intra- and extracellular compartments of the human host. *ΔompB* mutants show a low level of *vir*::*lacZ* expression which is not osmoregulated, but *envZ*::Tn*10* mutants show a low level of *β*-galactosidase activity which is still inducible under conditions of high osmolarity. This indicates that EnvZ is not the sole component involved in the response of *vir*::*lacZ* to osmolarity.

In *ΔompB* mutants, *ompC*, *ompF*, and *vir* genes are not able to respond to changes in osmolarity. The lack of expression of genes like *ompF* and *ompC* which are directly involved in the physiological response to changes in osmotic strength can affect the capacity of the bacteria to adapt to new conditions inside the host cell. In addition, the absence of two major components of the outer membrane could alter its complex structure. In this case, virulence gene products such as IpaB and IpaC could be prevented from inserting correctly into the outer membrane, and the invasive phenotype would not be fully expressed. On the other hand, the *vir* gene seems to belong to a class of genes whose expression is optimized in high-osmolarity conditions. In the *ΔompB* mutants, *vir* osmoregulation is abolished. It is perhaps for this reason that the bacteria cannot express the *vir* gene at the level required for the entry process.

4 Other Environmental Stimuli Affecting Virulence Gene Expression

4.1 Gene Induction by Congo Red Binding

A phenotype which is characteristic of invasive strains of *Shigella* is the ability to bind the dye Congo red (MAURELLI et al. 1984b; PAYNE and FINKELSTEIN 1977). These strains produce red pigmented colonies on agar medium containing Congo red and are called Crb^+ (Congo red binding) or Pcr^+ (pigmentation on Congo red plates). The ability of *Shigella* strains to bind Congo red correlates very closely with invasive ability and the presence of the virulence plasmid. The inability to bind Congo red is generally due to the complete loss of, or deletion within, the 220-kb virulence plasmid (MAURELLI et al. 1984b). As indicated earlier, the Congo red-binding phenotype is also temperature regulated in a fashion similar to regulation of virulence. Crb^+ bacteria are also able to bind hemin through a plasmid-encoded bacterial surface polypeptide of 101 kDa (STUGARD et al. 1989). However, this type of binding is independent of hemin transport since both Crb^+ and Crb^- strains can utilize hemin as a sole source of iron (DASKALEROS and PAYNE 1987). The presence of hemin prebound to the bacteria or HeLa cells greatly enhances the invasive ability of *Shigella* in tissue culture cells.

The exact relationship between Congo red binding and *Shigella* virulence is not known. Observations from a recent study (SANKARAN et al. 1989) suggest, however, that Congo red binding may mimic the binding and activity of a host cell factor encountered by intracellular bacteria which serves to regulate expression of *Shigella* membrane proteins. Virulent *S. flexneri*, *S. dysenteriae* serotype 1, and EIEC strains cultured in the presence of Congo red at 37 °C synthesize increased levels of membrane-associated proteins of 43, 58, and 63 kDa. These proteins are not detected in bacteria cultured at 30 °C even in the presence of Congo red, which suggests that temperature regulation of synthesis is dominant to induction by the dye. Bacteria grown in the absence of Congo red, however, express increased levels of the 43-kDa protein after invasion of tissue culture epithelial cells and guinea pig corneal epithelial cells. Amino acid sequence analysis of the 43-kDa protein perfectly matches the seqence of the *ipaC* gene. These data suggest that Congo red can induce expression of *Shigella* virulence genes. In addition, a host cell intracellular component may cause a similar temperature-dependent induction of protein expression in virulent strains of *Shigella*. The nature of the host cell factor which Congo red may be mimicking remains to be determined.

4.2 Influence of the Intracellular Compartment

It is generally accepted that the environment encountered by invasive bacteria within a host cell is different from that outside the host cell. *Yersinia*, in fact, respond to the differences in Ca^{2+} ion concentration between the extracellular and the intracellular compartment to regulate expression of certain virulence genes (STRALEY and BOWMER 1986). A recent report (HEADLEY and PAYNE 1990), describing the effects of intracellular and extracellular growth on protein expression by *S. flexneri*, suggests that *S. flexneri* serotype 2a differentially expresses proteins when exposed to these two different environments. In these studies, radiolabeled bacteria isolated from infected HeLa cell monolayers exhibit increased expression of proteins of 97, 62, 58, 50, 25, and 18 kDa, while proteins of 100, 85, 70, 64, and 55 kDa appear to be suppressed in their expression. The proteins suppressed during intracellular growth are, however, fully expressed by bacteria grown in tissue culture medium. Analysis of infected HeLa cell monolayers labeled at different times during the infectious cycle revealed that a bacterial protein of 58 kDa is induced during the invasion stage, while 62- and 25-kDa proteins are induced during intracellular multiplication. Radioimmunoprecipitation of bacterial proteins from intracellular and extracellular *S. flexneri*, with both convalescent monkey antisera and monoclonal antibodies specific for IpaB and IpaC, reveals that in the extracellular environment of tissue culture medium all of the Ipa proteins are synthesized. In contrast, IpaB and IpaC expression is undetectable in organisms isolated from the intracellular environment, while expression of an approximately 80-kDa protein corresponding in size to IpaA is induced. Additional proteins immunoprecipitated by the convalescent antisera include a protein of 140 kDa, corresponding to the size of the *icsA* gene product, a protein essential in the inter- and intracellular movement of invasive shigellae (BERNARDINI et al. 1989). These studies suggest that IpaA and IcsA may be induced in the bacteria during the intracellular stage of pathogenesis where they may be required as virulence factors.

There is a discrepancy in the levels of IpaC detected in the intracellular compartment in this report and the earlier study which examined induction by Congo red (SANKARAN et al. 1989). SANKARAN et al. (1989) found that a 43-kDa protein corresponding to IpaC increased after entry into HeLa cells. The reason for the contradictory observations is not clear. Both studies employed monoclonal antibodies to measure the levels of protein from bacteria isolated from infected HeLa cells. However, quantitation was performed by radioimmunoprecipitation in one study (HEADLEY and PAYNE 1990) and immunofluorescence in the other (SANKARAN et al. 1989). The relative sensitivity of IpaC to exogenous proteases may account for the different levels detected.

4.3 pH and Anaerobiosis

Shigellae are subjected to a dramatic range of pH changes during their transit through the alimentary canal and upper gastrointestinal tract to the colon. They therefore must be capable of adapting to different pH conditions in order to gain entry to the colonic epithelium. It seems reasonable to speculate that *Shigella* may employ specific regulatory systems for modulating gene expression according to pH changes it experiences in the environment. A strong capacity to resist the low pH of the stomach has apparently evolved in *Shigella* as the 50% infective dose (ID_{50}) is less than 500 organisms (DuPont et al. 1989). An analysis of the survival of *Salmonella*, *Shigella*, and EIEC in pH-adjusted media (pH1–7) and in gastric samples showed that *S. flexneri* is able to survive at pH 2, whereas *S. typhimurium* fails to survive below pH 3, and EIEC fail to survive below pH 4 (Gorden and Small 1990). These data reflect the differences in the inoculum size required for each pathogen to cause disease and suggest that differential survival in the low-pH environment of the stomach correlates with ID_{50}. The genetic basis of this low-pH survivability in *Shigella* is only now beginning to be studied, and, as specific genes are identified, their regulation in response to pH change can be studied.

Bacteria invading epithelial cells are also exposed to low pH in the phagocytic vacuole. The 58-kDa and 97-kDa proteins which are induced in the cytoplasmic compartment (see previous section) can also be induced in vitro. When *Shigella* are grown at pH 4.8 in minimal medium whose ion concentrations reflect intracellular levels, synthesis of these two proteins is increased (Headley and Payne 1990). This suggests that pH, perhaps in concert with specific ionic conditions, can regulate gene expression in *Shigella*. The existence of pH-regulated genes, however, would not be unexpected. Low pH has been shown to induce the expression of a subset of genes in *E. coli* K-12 and *S. typhimurium* (Slonczewski et al. 1987; Aliabadi et al. 1988). The assignment of a specific role for pH-regulated genes of *Shigella* in bacterial virulence is the next important step in this analysis.

Another environmental change which *Shigella* experiences in moving from the external world into the human host is oxygen limitation in the intestine. A role for anaerobic induction of virulence has been proposed in recent studies of *Salmonella* (Ernst et al. 1990; Lee and Falkow 1990; Schiemann and Shope 1991). These studies demonstrated that growth of *S. typhimurium* under anaerobic conditions leads to increased invasion in tissue culture cell models. Preliminary results suggest that anaerobic growth may produce an analogous effect in *Shigella* (Palchaudhuri, personal communication).

5 Other Regulators—*kcpA*

The *kcpA* locus was the first genetic locus determined to encode a gene essential for *Shigella* virulence (Formal et al. 1971). The phenotypic defect of the original *kcpA* mutant was described as an inability to provcke keratoconjunctivitis in a guinea pig eye (positive Sereny test). However, the mutant is still capable of penetrating HeLa cells in tissue culture. The *kcpA* locus maps to 12 min on the *Shigella* chromosome and shows 60% cotransductional linkage with *purE*. Experiments employing *E. coli-S. flexneri* hybrids carrying the *Shigella* invasion plasmid demonstrated a requirement for *kcpA* for full expression of the virulence phenotype. These hybrids are able to invade tissue culture cells, but are unable to produce a positive Sereny test unless the *kcpA* region of the *S. flexneri* chromosome is transferred in by conjugation or transduction (SANSONETTI et al. 1983). A more precise role for *kcpA* can be deduced from a comparison of the phenotypes of a *kcpA* mutant with a *virG* mutant. Both mutants are negative in the Sereny test, plaque assay negative, invasive for HeLa cells, but show no motility inside the cells and cannot spread from cell to cell. In addition, both mutants fail to express a 120–130-kDa protein which is the product of the *virG/icsA* gene and is required for intracellular motility of the bacteria as well as intercellular spread (BERNARDINI et al. 1989; MAKINO et al. 1986; PAL et al. 1989). Therefore, it appears that expression of *virG/icsA* is dependent on a functional *kcpA*. Activation of *virG/icsA* transcription by *kcpA* could explain the similar phenotypes seen in *kcpA* and *virG/icsA* mutants. Another activator of *virG*, *virF*, was described in an earlier section. It is not clear whether *virF* and *kcpA* act together or independently to affect expression of *virG/icsA*.

Spontaneous $KcpA^+$ variants of *E. coli* K-12 can be isolated which behave exactly like the hybrids described above which had inherited the *kcpA* region from a *S. flexneri* donor (PAL et al. 1989). This suggests that an analogue of *kcpA* exists in *E. coli* K-12, but is not expressed.

The *kcpA* gene has not yet been cloned. An earlier report of the cloning and sequencing of *kcpA* proved to be incorrect (YAMADA et al. 1989). The clone, in fact, was a DNA fragment carrying *virR*, and the reported sequence for *kcpA* matched the sequence of *virR/hns* (HULTON et al. 1990). This indicates that one regulatory gene (*virR*) cloned on a multicopy number plasmid (pBR322 based) may suppress the effect of a mutation in another regulatory gene (*kcpA*). However, this conclusion presents a paradox which is not easily resolved: how can *virR* mediate this phenotypic suppression since *virR* acts as a repressor of virulence gene expression at 30 °C, while *kcpA* appears to be a positive regulator of *virG*? Further characterization of the nature of the mutant *kcpA* locus as well as identification of the *kcpA* product should help resolve this paradox.

6 Perspectives

Both positive and negative regulatory mechanisms are employed in *Shigella* virulence gene regulation, and the genes for these regulators are found on the chromosome as well as on the virulence plasmid. Environmental stimuli other than temperature also serve to regulate virulence gene expression.

The positive activators *virF* and *virB* probably interact in some way with the *virR/hns* gene product in mediating temperature control of virulence gene expression. VirR/H-NS can be viewed as a "master switch" which somehow senses the change in temperature, perhaps in conjunction with the ribosome or another gene product. VirF and VirB can then activate expression of the virulence regulon and produce the virulent phenotype. Since temperature appears to be dominant to the other environmental stimuli, it is perhaps at the level of VirF and VirB that the other stimuli affect gene expression. In this fashion, the positive activators may respond to osmolarity or oxygen limitation in the host environment by "fine-tuning" expression of the already activated virulence regulon. Thus, a two-tiered system of regulation would permit amplification (or suppression) of certain key genes at the permissive temperature in the appropriate host compartment. It is evident that much more can be learned by further study of *virR/hns* and its relationship with *virF* and *virB*. A determination of potential DNA-protein binding interactions between these regulators and promoter regions of known virulence genes is particularly important as well.

Analogues of two regulators of *Shigella* virulence, *virR/hns and ompB*, are found in *E. coli* K-12. A mutated form of a third regulator, *kcpA*, also exists in *E. coli* K-12. This may not be too surprising since *Shigella* and *E. coli* are so closely related at the DNA level that they should be considered two species of the same genus. It is known that these chromosomal regulatory genes serve other non-virulence-related functions in these organisms. On the other hand, the G+C content of the virulence genes of the 220-kb virulence plasmid is very low relative to the G+C content of the *Shigella* and *E. coli* chromosomes. It has been proposed that the virulence plasmid was acquired by *Shigella* from an unrelated organism outside the genus (BAUDRY et al. 1988). The virulence regulon of this "foreign" plasmid has apparently adopted these chromosomal regulatory genes and integrated them into the sensing/signaling system *Shigella* uses to regulate the virulence phenotype.

The mammalian cytoplasm presents a microenvironment with multiple stimuli which the bacterium can use to modulate expression of its genes. The study of gene induction in the intracellular compartment is another exciting area of investigation in *Shigella* virulence gene regulation. It is likely that efforts here will help delineate the roles of nontemperature effectors of *Shigella* virulence such as pH, osmolarity, and anaerobiosis.

In summary, the pathogenicity of *Shigella* is multigenic, both from the point of view of the structural genes required for invasion and intercellular spread, as well as the genes which regulate expression of these virulence determinants. The complexity of virulence gene regulation in *Shigella* is only now beginning to be fully appreciated. Future studies in this area will no doubt provide both a better understanding of virulence gene regulation as well as an insight into the mechanisms of *Shigella* pathogenicity.

Acknowledgements. Research on the genetics of *Shigella* virulence in the laboratory of A.T.M. is currently supported by USUHS protocol RO-7385 and Public Health Service grant AI24656 from the National Institute of Allergy and Infectious Diseases. Thanks to Dan Rowley for his careful reading of this manuscript.

References

Abeles AL, Freidman SA, Austin SJ (1985) Partition of unit-copy miniplasmids to daughter cells. III. The DNA sequence and functional organization of the P1 partition region. J Mol Biol 185: 261–272

Adler B, Sasakawa C, Tobe T, Makino S, Komatsu K, Yoshikawa M (1989) A dual transcriptional activation system for the 230 kb plasmid genes coding for virulence-associated antigens of *Shigella flexneri*. Mol Microbiol 3: 627–635

Aliabadi Z, Park YK, Slonczewski JL, Foster JW (1988) Novel regulatory loci controlling oxygen- and pH-regulated gene expression in *Salmonella typhimurium*. J Bacteriol 170: 842–851

Andrews GP, Hromockyj AE, Coker C, Maurelli AT (1991) Two novel virulence loci, *mxiA* and *mxiB*, in *Shigella flexneri* 2a facilitate excretion of invasion plasmid antigens. Infect Immun 59: 1997–2005

Arico B, Miller FJ, Craig R, Stibitz S, Monack D, Falkow S, Gross R, Rappuoli R (1989) Sequences required for expression of *Bordetella pertussis* virulence factor share homology with prokaryotic signal transduction proteins. Proc Natl Acad Sci USA 86: 6671–6675

Baudry B, Kaczorek M, Sansonetti PJ (1988) Nucleotide sequence of the invasion plasmid antigen B and C genes (*ipaB* and *ipaC*). Microb Pathog 4: 345–357

Bernardini ML, Fontaine A, Sansonetti PJ (1990) The two-component regulatory system OmpR-EnvZ controls the virulence of *Shigella flexneri*. J Bacteriol 172: 6274–6281

Bernardini ML, Mounier J, d'Hauteville H, Coquis-Rondon M, Sansonetti PJ (1989) Identification of *icsA*, a plasmid locus of *Shigella flexneri* that governs bacterial intra- and intercellular spread through interaction with F-actin. Proc Natl Acad Sci USA 86: 3867–3871

Braun G, Cole ST (1982) The nucleotide sequence coding for major outer membrane protein OmpA of *Shigella dysenteriae*. Nucleic Acids Res 10: 2367–2378

Brenner DJ, Fanning GR, Sherman FJ, Falkow S (1972) Polynucleotide divergence among strains of *Escherichia coli* and closely related organisms. J Bacteriol 109: 953–965

Buysse JM, Venkatesan M, Mills JA, Oaks EV (1990) Molecular characterization of a trans-acting, positive effector (*ipaR*) of invasion plasmid antigen synthesis in *Shigella flexneri* serotype 5. Microb Pathog 8: 197–211

Comeau DEK, Ikenaka K, Tsung K, Inouye M (1985) Primary characterization of the protein products of the *Escherichia coli ompB* locus: structure and regulation of synthesis of the OmpR and EnvZ proteins. J Bacteriol 164: 578–584

Cossart P, Groisman EA, Serre M, Casadaban MJ, Gicquel-Sanzey B (1986) *crp* genes of *Shigella flexneri*, *Salmonella typhimurium*, and *Escherichia coli*. J Bacteriol167: 639–646

Csonka LN (1989) Physiological and genetic responses of bacteria to osmotic stress. Microbiol Rev 56: 121–147

Daskaleros PA, Payne SM (1987) Congo red binding phenotype is associated with hemin binding and increased infectivity of *Shigella flexneri* in the HeLa cell model. Infect Immun 55: 1393–1398

Deretic V, Dikshit R, Konyescni WM, Chakrabarty AM, Misra TK (1989) The *algR* gene, which regulates mucoidy in *Pseudomonas aeruginosa*, belongs to a class of environmentally responsive genes. J Bacteriol 171: 1278–1283

DiRita JV, Mekalanos JJ (1989) Genetic regulation of bacterial virulence. Annu Rev Genet 23: 455-482

Dorman CJ, NiBhriain N, Higgins CF (1990) DNA supercoiling and environmental regulation of virulence gene expression in *Shigella flexneri*. Nature 344: 789–792

Drlica K, Rouviere-Yaniv J (1987) Histone like proteins of bacteria. Microbiol Rev 51: 301–319

DuPont HL, Levine MM, Hornick RB, Formal SB (1989) Inoculum size in shigellosis and implications for expected mode of transmission. J Infect Dis 159: 1126–1128

Ernst RK, Dombroski DM, Merrick JM (1990) Anaerobiosis, type 1 fimbriae, and growth phase are factors that affect invasion of HEp-2 cells by *Salmonella typhimurium*. Infect Immun 58: 2014–2016

Falconi M, Gualtieri MT, LaTeana A, Lasso MA, Pon CL (1988) Proteins from the prokaryotic nucleoid: primary and quaternary structure of the 15 kD *Escherichia coli* DNA binding protein H-NS. Mol Microbiol 2: 323–329

Formal SB, Gemski P Jr, Baron LS, LaBrec EH (1971) A chromosomal locus which controls the ability of *Shigella flexneri* to evoke keratoconjunctivitis. Infect Immun 3: 73–79

Forst S, Comeau D, Shigemi D, Inouye M (1987) Localization and membrane topology of EnvZ, a protein involved in osmoregulation of OmpF and OmpC in *Escherichia coli.* J Biol Chem 262: 16433–16438

Forst S, Delgado J, Inouye M (1989) Phosphorylation of OmpR by the osmosensor EnvZ modulates expression of the *ompF* and *ompC* genes in *Escherichia coli.* Proc Natl Acad Sci USA 86: 6052–6056

Friedrich K, Gualerzi CO, Lammi M, Pon CL (1988) Proteins from the prokaryotic nucleoid: interaction of nucleic acids with the 15 kDa *Escherichia coli* histone-like protein H-NS. FEBS Lett 229: 197–202

Galan JE, Curtiss R III (1990) Expression of *Salmonella typhimurium* genes required for invasion is regulated by changes in DNA supercoiling. Infect Immun 58: 1879–1885

Goodman HM, Abelson J, Landy A, Brenner S, Smith JD (1968) Amber suppression: a nucleotide change in the anticodon of a tyrosine transfer RNA. Nature 217: 1019–1024

Göransson M, Uhlin BE (1984) Environmental temperature regulates transcription of a virulence pili operon in *E. coli.* EMBO J 3: 2885–2888

Göransson M, Sondén B, Nilsson P, Dagberg B, Forsman K, Emanuelson K, Uhlin BE (1990) Transcriptional silencing and thermoregulation of gene expression in *Escherichia coli.* Nature 344: 682–685

Gorden J, Small P (1990) Quantitative analysis of the effects of variable pH on the growth of *Shigella flexneri*, *Salmonella typhimurium*, and enteroinvasive *Escherichia coli.* Abstracts of the 90th annual meeting of the American Society for Microbiology, Washington, p 64

Gottesman S (1984) Bacterial regulation: global regulatory networks. Annu Rev Genet 18: 415–441

Gualerzi CO, Losso MA, Lammi M, Friedrich K, Pawlik RT, Canonaco MA, Gianfranceschi G, Pingoud A, Pon CL (1986) Proteins from the prokaryotic nucleoid. Structural and function characterization of the *Escherichia coli* DNA-binding proteins NS (HU) and H-NS. In: Gualerzi CO, Pon CL (eds) Bacterial chromatin. Springer, Berlin Heidelberg New York, pp 101–134

Hancock REW (1987) Role of porins in other membrane permeability. J Bacteriol 163: 929–933

Headley VL, Payne SM (1990) Differential protein expression by *Shigella flexneri* in intracellular and extracellular environments. Proc Natl Acad Sci USA 87: 4179–4183

Higgins CF, Dorman CJ, NiBhriain N (1990) Environmental influences on DNA supercoiling: a novel mechanism for the regulation of gene expression. In: Drlica K, Riley M (eds) The bacterial chromosome. American Society for Microbiology, Washington, D.C., pp 421–434

Hromockyj AE, Maurelli AT (1989a) Identification of *Shigella* invasion genes by isolation of temperature-regulated *inv*: : *lacZ* operon fusions. Infect Immun 57: 2963–2970

Hromockyj AE, Maurelli AT (1989b) Identification of an *Escherichia coli* gene homologous to *virR*, a regulator of *Shigella* virulence. J Bacteriol 171: 2879–2881

Hulton CSJ, Seirafi A, Hinton JCD, Sidebotham JM, Waddell L, Pavitt GD, Owen-Hughes T, Spassky A, Buc H, Higgins CF (1990) Histone-like protein HI (H-NS), DNA supercoiling, and gene expression in bacteria. Cell 63: 631–642

Jones PG, VanBogelen RA, Neidhardt FC (1987) Induction of proteins in response to low temperature in *Escherichia coli.* J Bacteriol 169: 2092–2095

Kato J-I, Ito K-I, Nakamura A, Watanabe H (1989) Cloning of regions required for contact hemolysis and entry into LLC-MK2 cells from *Shigella sonnei* Form I plasmid: *virF* is a positive regulator gene for these phenotypes. Infect Immun 57: 1391–1398

Kawula T, Orndorff P (1991) Rapid site specific DNA inversion in *Escherichia coli* mutants lacking the histone-like protein H-NS. J Bacteriol 173: 4116–4123

LaTeana A, Falconi M, Scarlato V, Lammi M, Pon CL (1989) Characterization of the structural genes for the DNA-binding protein H-NS in Enterobacteriaceae. FEBS Lett 244: 34–38

Lee CA, Falkow S (1990) The ability of *Salmonella* to enter mammalian cells is affected by bacterial growth state. Proc Natl Acad Sci USA 87: 4304–4308

Leroux B, Yanofsky MF, Winas SC, Ward JE, Ziegler SF, Nester EW (1987) Characterization of the *virA* locus of *Agrobacterium tumefaciens*: a transcriptional regulator and host range determinant. EMBO J 6: 849–856

Lugtenberg B, Peters R, Bernheimer M, Berendson W (1976) Influence of cultural conditions and mutations on the composition of the outer membrane proteins of *Escherichia coli.* Mol Gen Genet 147: 251–262

Makino S, Sasakawa C, Kamata K, Kurata T, Yoshikawa M (1986) A genetic determinant required for continuous reinfection of adjacent cells on large plasmid in *S. flexneri* 2a. Cell 46: 551–555

Marsh M, Hillyard DR (1990) Nucleotide sequence of *hns* encoding the DNA-binding protein H-NS of *Salmonella typhimurium*. Nucleic Acids Res 18: 3397

Maurelli AT (1989a) Regulation of virulence genes in *Shigella*. Mol Biol Med 6: 425–432

Maurelli AT (1989b) Temperature regulation of virulence genes in pathogenic bacteria: a global strategy for human pathogens. Microb Pathog 7: 1–10

Maurelli AT, Curtiss R III (1984) Bacteriophage Mu*d*I (Ap^r *lac*) generates *vir-lac* operon fusions in *Shigella flexneri* 2a. Infect Immun 45: 642–648

Maurelli AT, Sansonetti PJ (1988) Identification of a chromosomal gene controlling temperature-regulated expression of *Shigella* virulence. Proc Natl Acad Sci USA 85: 2820–2824

Maurelli AT, Blackmon B, Curtiss R III (1984a) Temperature-dependent expression of virulence genes in *Shigella* species. Infect Immun 43: 195–201

Maurelli AT, Blackmon B, Curtiss R III (1984b) Loss of pigmentation in *Shigella flexneri* 2a is correlated with loss of virulence and virulence-associated plasmid. Infect Immun 43: 397–401

May G, Dersch P, Haardt M, Middendorf A, Bremer E (1990) The *osmZ* (*bglY*) gene encodes the DNA-binding protein H-NS (Hla), a component of the *Escherichia coli* K12 nucleoid. Mol Gen Genet 224: 81–90

Miller JF, Mekalanos JJ, Falkow S (1989) Coordinate regulation and sensory transduction in the control of bacterial virulence. Science 243: 916–922

Miller SI, Kukral AM, Mekalanos JJ (1989) A two-component regulatory system (*phoP-phoQ*) controls *Salmonella typhimurium* virulence. Proc Natl Acad Sci USA 86: 5054–5058

Mori H, Kondo A, Oshima T, Hiraga S (1986) Structure and function of the F plasmid genes essential for partitioning. J Mol Biol 192: 1–15

Neidhardt FC, VanBogelen RA (1987) Heat shock response. In: Neidhardt FC (ed) *Escherichia coli* and *Salmonella typhimurium* cellular and molecular biology. American Society for Microbiology, Washington, D C, pp 1334–1345

Nikaido H, Vaara M (1985) Molecular basis of bacterial outer membrane permeability. Microbiol Rev 49: 1–32

Norioka S, Ramakrishnan R, Ikenaka K, Inouye M (1986) Interaction of a transcriptional activator, OmpR, with reciprocally osmoregulated genes *ompF* and *ompC* of *Escherichia coli*. J Biol Chem 257: 13692–13698

Ochman H. Whittam TS, Caugant DA, Selander RK (1983) Enzyme polymorphism and genetic population structure in *Escherichia coli* and *Shigella*. J Gen Microbiol 129: 2715–2726

Pal T, Newland JW, Tall BD, Formal SB, Hale TL (1989) Intracellular spread of *Shigella flexneri* associated with the *kcpA* locus and a 140-kilodalton protein. Infect Immun 57: 477–486

Payne SM, Finkelstein RA (1977) Detection and differentiation of iron-responsive avirulent mutants on Congo red agar. Infect Immun 18: 94–98

Pon CL, Calogero RA, Gualerzi CO (1988) Identification, cloning, nucleotide sequence and chromosomal map location of *hns*, the structural gene for *Escherichia coli* DNA-binding protein H-NS. Mol Gen Genet 212: 199–202

Ronson CW, Nixon BT, Ausubel FM (1987) Conserved domains in bacterial regulatory proteins that respond to environmental stimuli. Cell 49: 579–581

Sakai T, Sasakawa C, Makino C, Kamata K, Yoshikawa M (1986a) Molecular cloning of a genetic determinant for Congo red binding ability which is essential for the virulence of *Shigella flexneri*. Infect Immun 51: 476–482

Sakai T, Sasakawa C, Makino C, Yoshikawa M (1986b) DNA sequence and product analysis of the *virF* locus responsible for Congo red binding and cell invasion in *S. flexneri* 2a. Infect Immun 54: 395–402

Sakai T, Sasakawa C, Yoshikawa M (1988) Expression of four virulence antigens of *Shigella flexneri* is positively regulated at the transcriptional level by the 30 kilodalton *virF* protein. Mol Microbiol 2: 589–597

Sankaran K, Ramachandran V, Subrahmanyam YVBK, Rajarathnam S, Elango S, Roy RK (1989) Congo red-mediated regulation of levels of *Shigella flexneri* 2a membrane proteins. Infect Immun 57: 2364–2371

Sansonetti PJ, Hale TL, Dammin GJ, Kapfer C, Collins HH Jr, Formal SB (1983) Alterations in the pathogenicity of *Escherichia coli* K-12 after transfer of plasmid and chromosomal genes from *Shigella flexneri*. Infect Immun 39: 1392–1402

Sasakawa C, Makino S, Kamata K, Yoshikawa M (1986) Isolation, characterization, and mapping of Tn5 insertions in the 140 megadalton invasion plasmid defective in the mouse Séreny test in *Shigella flexneri* 2a. Infect Immun 54: 32–36

Schiemann DA, Shope SR (1991) Anaerobic growth of *Salmonella typhimurium* results in increased uptake by Henle 407 epithelial and mouse peritoneal cells in vitro and repression of a major outer membrane protein. Infect immun 59: 437–440

Sekiya T, Gait MJ, Noris K, Ramamoorthy B, Khorana HG (1976) The nucleotide sequence in the promoter region of the gene for an *Escherichia coli* tyrosine transfer ribonucleic acid. J Biol Chem 251: 5124–5140

Slonczewski JL, Gonzalez TN, Bartholomew FM, Holt NJ (1987) Mu d-directed *lacZ* fusions regulated by low pH in *Escherichia coli*. J Bacteriol 169: 3001–3006

Spassky A, Buc HD (1977) Physio-chemical properties of a DNA binding protein: *Escherichia coli* factor H_1. Eur J Biochem 81: 79–90

Spassky A, Timsky S, Garreau H, Buc H (1984) Hla, an *E. coli* DNA-binding protein which accumulates in stationary phase, strongly compacts DNA in vitro. Nucleic Acids Res 12: 5321–5340

Straley SC, Bowmer WS (1986) Virulence genes regulated at the transcriptional level by Ca^{2+} in *Yersinia pestis* include structural genes for outer membrane proteins. Infect Immun 51: 445–454

Stock JB, Ninfa AJ, Stock A (1989) Protein phosphorylation and regulation of adaptive responses in bacteria. Microbiol Rev 53: 450–490

Stugard CE, Daskaleros PA, Payne SM (1989) A 101-kilodalton heme-binding protein associated with Congo red binding virulence of *Shigella flexneri* and enteroinvasive *Escherichia coli* strains. Infect Immun 57: 3534–3539

Tobe T, Nagai S, Okada N, Adler B, Yoshikawa M, Sasakawa C (1991) Temperature-regulated expression of invasion genes in *Shigella flexneri* is controlled through the transcriptional activation of the *virB* gene on the large plasmid. Mol Microbiol 5: 887-893

VanBogelen RA, Neidhardt FC (1990) Ribosomes as sensors of heat and cold shock in *Escherichia coli*. Proc Natl Acad Sci USA 87: 5589–5593

VanBogelen RA, Hutton ME, Neidhardt FC (1990) Gene-protein database of *Escherichia coli* K-12: edition 3. Electrophoresis 11: 1131–1166

Watanabe H, Arakawa E, Ito K-I, Kato J-I, Nakamura A (1990) Genetic analysis of an invasion region by use of a Tn*3-lac* transposon and identification of a second positive regulator gene, *invE*, for cell invasion of *Shigella sonnei*: significant homology of InvE with ParB of plasmid P1. J Bacteriol 172: 619–629

Yamada M, Sasakawa C, Okada N, Makino S-I, Yoshikawa M (1989) Molecular cloning and characterization of chromosomal virulence region *kcpA* of *Shigella flexneri*, Mol Microbiol 3: 207-213

Pathogenesis and Immunology in Shigellosis: Applications for Vaccine Development*

T. L. HALE[1] and D. F. KEREN[2]

1 Protective Immunity in Shigellosis

Although the immunological basis of protection from shigellosis is poorly understood, it is likely that secretory immunoglobulin A (IgA) on mucosal surfaces and mucosal lymphocytes in the intestine play important roles in this process. Therefore, a summary of the possible protective roles of lymphocytes and immunoglobulin will provide a context for the subsequent discussion of *Shigella* vaccine development.

1.1 The Role of Gut-Associated Lymphoid Tissues in Antigen Sampling and in Disease Pathology

Specific lymphoid tissues in the intestine play a key role in the mucosal immune response. For example, Peyer's patches, isolated lymphoid follicles, and the

* The views of the authors do not purport to reflect the position of the Department of the Army or the Department of Defense (paragraph 4-3, AR 360-5)

[1] Department of Enteric Infections, Walter Reed Army Institute of Research, Washington, DC 20307-5100, USA
[2] Ward Medical Laboratories, Ann Arbor, MI 48105, USA

Current Topics in Microbiology and Immunology, Vol. 180

appendix are component structures of the gut-associated lymphoid tissues (GALT). Since mesenteric lymph nodes (MLN) also contain precursor lymphoid cells, they can be included as part of the GALT despite their less intimate association with the gut lumen. These nodes probably serve as waystations for lymphocytes passing from the intestine to the systemic circulation. Peyer's patches are grossly identifiable aggregates of lymphoid nodules that occur on the antimesenteric border of the small intestine, most prominently in the terminal ileum. Most in vitro studies on mucosal lymphocytes have been performed using cells from the Peyer's patch. In experimental animals, Peyer's patches can be detected about halfway through gestation, and the lymphoid tissue in the patch proliferates rapidly in the fetus. At birth, the patches have the greatest density of all proliferating lymphoid cells in the body. The increase in size and number of these structures after birth reflects the initial response of the mucosal immune system to the environmental antigens, especially the micronial flora. For example, germ-free animals have poorly developed Peyer's patches that enlarge only after colonization of the intestine with microorganisms.

The dome-corona area of the Peyer's patch, located just above the follicles and beneath the surface epithelium, contains many immune region-associated antigen (Ia) class II-positive macrophages. These phagocytic cells process and present antigen to T helper cells that are activated to produce lymphokines. The large follicular areas of Peyer's patches are rich in B lymphocytes that mainly express surface IgM and serve as precursors for IgA-secreting plasma cells (TSENG 1984). In addition to the organized Peyer's patches, isolated lymphoid follicles are found throughout the gastrointestinal tract. In higher primates, these isolated follicles are the most abundant discrete lymphoid structures of the GALT. The overlying surface epithelium of these follicles is similar to the follicle-associated epithelium (FAE) that overlies Peyer's patches. Since extirpation or exclusion of Peyer's patches does not preclude mucosal immune responses in experimental animals, the isolated lymphoid follicles appear to function as antigen sampling and processing centers which supplement Peyer's patches (KEREN et al. 1978).

The specialized surface epithelial cells that serve as a portal of entry for intraluminal antigens have been termed M (membranous) cells. These cells are located within the FAE of Peyer's patches and in isolated lymphoid follicles in the intestine, tonsils, and appendix (ROSNER and KEREN 1984). The M cell designation describes the flat appearance and broad irregular microvilli that differentiate these cells from adjacent absorptive epithelial cells (OWEN 1977). When specific pathogen-free mice are transferred to conventional animal housing, a threefold increase in M cells is measurable after 1 week (SMITH et al. 1987). This proliferation of M cells facilitates the uptake of viruses, bacteria, and protozoa by the GALT. The mechanism by which the particles attach to M cells is unclear. The mucous layer is thinner over M cells than over absorptive epithelial cells, and the glycocalyx of M cells may also be thinner. Therefore M cells may be inherently "stickier" than the absorptive epithelium (SNELLER and STROBER 1986).

M cells may play a key role in the pathogenesis of shigellosis since *Shigella flexneri* preferentially colonize the FAE of ligated rabbit ileal loops at an early time point (90 min) after injection of the organisms into the intestinal lumen (Fig. 1). At this time, virulent strains of shigellae show evidence of intracellular multiplication within the M cells, and after 18 h. ulcerations are observed in the domed regions over Peyer's patches (WASSEF et al. 1989). It is also possible that virulent shigellae spread laterally to adjacent absorptive epithelial cells from the infected FAE, and this intercellular spread could result

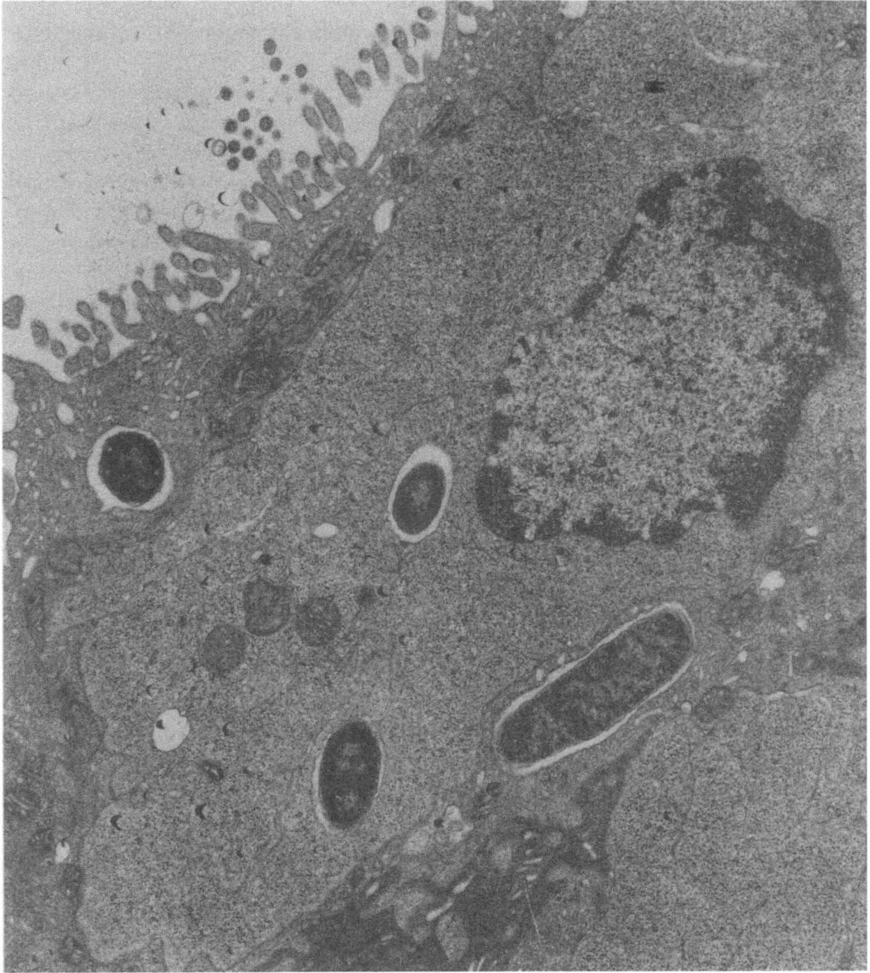

Fig. 1. Transmission electron micrograph of rabbit M cells containing *S. flexneri* 2a within endocytic vacuoles. × 12 000

in at least some of the mucosal lesions that characterize shigellosis. In addition to pathogenic manifestations, the uptake of shigellae by M cells presumably facilitates the translocation of these organisms to antigen-processing cells in the lymphoid follicle. The balance between bacterial growth within the FAE and the immunopotentiated killing of these organisms within the underlying lymphoid follicle may well be a determining factor in the initial development of clinical disease.

Lymphocytes with specific functions selectively populate defined anatomical compartments of the intestinal mucosa. Interepithelial lymphocytes (IEL) and lamina propria lymphocytes (LPL) are distinct populations with vastly different functions. For example, IEL within the absorptive epithelium of the small intestine and colon consist mainly of T cells expressing surface antigens associated with suppressor/cytotoxic functions (CD8), while the LPL consist predominantly of CD4 T cells associated with helper functions (TREJDOSIEWICZ et al. 1987). Nonetheless, both IEL and LPL evidence antibody-dependent, cell-mediated cytotoxicity (ADCC) activity against *S. flexneri*, and IgA recognizing the *Shigella* somatic antigen is more effective than IgG in arming these lymphocytes (TAGLIABUE et al. 1984). In contrast, IgG and IgM are more effective than IgA in opsonizing shigellae for phagocytosis by polymorphonuclear neutrophils (REED 1975). Thus, interstitial IgG and IgM recognizing the *Shigella* somatic antigen may play an important role in eliminating shigallae that are released from degenerating absorptive epithelial cells into the lamina propria. It is also possible that Ia class II-restricted CD4 cytotoxic T cells recognize absorptive epithelial cells that have been infected by shigellae, but a role for cytotoxic T cells in eliminating infected enterocytes from the intestinal epithelium has not yet been demonstrated.

1.2 Secretory IgA and the Mucosal Immune Response

After antigenic stimulation in the GALT, antigen-specific B lymphocytes leave the mucosa and travel through the mesenteric lymph nodes, the thoracic duct, the spleen, and eventually back to the lamina propria of the gut and other mucosal locations (ROUX et al. 1981). This migration of B cells has led to the concept of a common mucosal immune system whereby all mucosal surfaces are primed by oral antigen administration. In addition, the brief tenure of antibody-secreting B cells in the peripheral circulation has allowed the sampling of immune response resulting from mucosal stimulation using the enzyme-linked immunospot (ELISPOT) assay (CZERKINSKY et al. 1983; HERRINGTON et al. 1990). The IgA-producing B cells are the most prominent mucosal immunocytes with 500 cells/mm^2 of colonic mucosa as compared to fewer than 50 IgM and IgG cells/mm^2 (BRANDTZAEG 1989).

Chronically isolated rabbit ileal (Thiry-Vella) loops have been used to study the development of secretory IgA responses against *S. flexneri* (KEREN et al. 1978). In this procedure, a 20-cm segment of ileun is isolated along with the intact vascular supply, and this loop can be directly stimulated by irrigating the lumen with antigen. Since approximately 2 ml mucosal secretion can be collected daily, the kinetics and specificity of mucosal antibody response can also be monitored. Direct stimulation of isolated loops with either invasive or noninvasive strains of *Shigella* elicits strong local secretory IgA responses against the somatic antigen of the challenge strain. Inclusion of a Peyer's patch within the isolated loop enhances this IgA response. In contrast to IgA, little IgG recognizing the somatic antigen is detected in the mucosal secretions, and direct stimulation of the isolated loops elicits little or no serum IgG recognizing the somatic antigen.

A secretory IgA memory response can also be demonstrated in rabbit ileal loops following peroral priming with *S. flexneri*. This was illustrated in experiments using rabbits that were primed with three weekly doses of an invasive *Escherichia coli–S. flexneri* 2a hybrid that does not elicit mucosal lesions. Two months after the third dose, chronically isolated loops were surgically created, and the animals were given an oral booster dose of the *Shigella* hybrid strain. Within 4 days, the primed animals demonstrated a significant memory response of secretory IgA recognizing the serotype 2a somatic antigen (KEREN 1983). Similar results were obtained after peroral priming with a noninvasive strain of *S. flexneri* that lacks the *Shigella* virulence plasmid (KEREN et al. 1983).

The major action of secretory IgA in protecting against most enteric infections is to neutralize toxic products and/or to prevent the attachment of microorganisms to the mucosa (FUBARA and FRETER 1973). However, secretory IgA collected from ileal loops of rabbits immunized with the invasive *E. coli–S. flexneri* 2a hybrid failed to inhibit the invasion of cultured mammalian cells by *S. flexneri* 2a (T.L. HALE and D.F. KEREN, unpublished observation). Antiserum elicited by parenteral immunization of rabbits with heat-killed *Shigella* also fails to inhibit the invasion of tissue culture cells (HALE and BONVENTRE 1979). Thus, antibody recognizing the lipopolysaccharide (LPS) somatic antigen may be ineffectual in inhibiting the initial uptake of shigellae by the intestinal epithelium. In addition to LPS, the plasmid-encoded invasion plasmid antigens (Ipa) proteins elicit strong IgG and IgA mucosal immune responses after *Shigella* infection (OAKS et al. 1986; DINARI et al. 1987). The Shiga cytotoxin/enterotoxin of *S. dysenteriae* 1 also elicits strong IgA responses (KEREN et al. 1989). A protective role of antibodies recognizing these proteinaceous virulence determinants has not yet been demonstrated.

2 Protective Immunity Evoked by *Shigella* Infection

Epidemiological observations suggest that *Shigella* infections elicit serotype-specific immune protection. For example, the incidence of clinical shigellosis peaks at 1–4 years and decreases to approximately 25% of peak levels in older children and adults living in endemic areas (BENNISH et al. 1990). Although the lower incidence of disease in the latter individuals is a result of a combination of factors such as mature standards of personal hygiene and improved nutrition, active immunity induced by environmental exposure to shigellae is probably the key parameter. The importance of immunity induced by environmental exposure is illustrated by the high attack rates of shigellosis observed in individuals newly arrived in endemic areas (DUPONT et al. 1970). Individuals with evidence of preexisting serotype-specific serum antibody also have significantly lower attack rates during *Shigella* outbreaks (COHEN et al. 1988). However, the occurrence of epidemics of *S. dysenteriae* 1 in areas endemic to other *Shigella* serotypes (GANGAROSA et al. 1969) suggests that the immunity induced by environmental exposure can be circumvented by the introduction of a serologically heterologous strain.

Serotype-specific protection has been demonstrated experimentally in rhesus monkeys that were orally challenged with *S. flexneri* 2a and then rechallenged with either *S. flexneri* or with *S. sonnei*. The animals that were rechallenged with the homologous strain were fully protected from disease; those challenged with the heterologous serotype experienced the same attack rate as naive controls (S.B. FORMAL et al. 1991). Immune protection was also demonstrated in volunteers that were infected previously with *S. flexneri* (DUPONT et al. 1972) or with *S. sonnei* (HERRINGTON et al. 1990). Although the number of volunteers was small, 50%–75% protection against experimental rechallenge was observed in these human studies. Since the volunteers were rechallenged fewer than 12 months after the initial shigella infection, the duration of immune protection is unknown.

3 Living Oral *Shigella* Vaccines

Until approximately 25 years ago, experimental *Shigella* vaccines consisted of heat- or acetone-killed whole cell preparations that were injected subcutaneously. In one particularly discouraging Egyptian field, administration of a total of 2.5×10^9 heat-killed organisms divided into three weekly doses had no effect upon the 25% attack rate of the homologous *S. flexneri* 3 serotype (HIGGINS et al. 1955). In later experiments using orally challenged rhesus monkeys, four injections of heat-killed *S. flexneri* 2a followed by a injection of acetone-killed vaccine elicited no protection against the homologous serotype. Indeed, even subcutaneous injection of living *S. flexneri* 2a failed to protect monkeys from oral challenge (FORMAL et al. 1967). When administered orally in five doses of 30 mg each, acetone-killed and dried whole bacterial cells expressing the *S. flexneri* 2a somatic antigen also failed to protect rhesus monkeys from a sunsequent *S. flexneri* 2a challenge (FORMAL et al. 1965). Owing to the apparent lack of efficacy of killed bacterial vaccines, *Shigella* vaccine research in the last 25 years has concentrated on the development of living, attenuated strains that will elicit the protective immune response of a *Shigella* infection without inducing the symptomatology of shigellosis. These vaccines can be broadly characterized as (a) *Shigella* strains with attenuating chromosomal mutations; (b) *Shigella* strains with attenuating mutations in the virulence plasmid; and (c) hybrid *Shigella* or *E. coli* strains expressing an incomplete complement of *Shigella* virulence determinants.

3.1 *Shigella* Vaccines with Attenuating Chromosomal Mutations

Table 1 lists some attributes of *Shigella* vaccines that are attenuated as a result of either spontaneous chromosomal mutations (SmD), a combination of spontaneous mutation and chemical mutation (Pur^-/Rif), spontaneous mutation and ultraviolet mutagenesis (TSF-21), or insertional mutation (Sfl-114). The earliest of these vaccines, streptomycin-dependent (SmD) strains of *S. flexneri*, were derived from spontaneous streptomycin-resistant (SmR) mutants that occur at a frequency of 10^{-8} when *S. flexneri* are inoculated onto mutrient agar plates containing streptomycin 400 μg/ml. Screening of SmR colonies for the inability to grow in the absence of the antibiotic allowed the isolation of SmD colonies at a frequency of 10^{-2}–10^{-3} (MEL et al. 1965a). The SmR mutations are located within the *rspL* ribosomal subunit gene, and the SmD phenotype is probably the result of an additional mutation changing the conformation of the 30-kDa ribosomal protein encoded by this gene. Although SmD strains cannot grow in the absence of streptomycin, they are shed from the intestine for approximately 3 days when given at doses of greater than 5×10^{10} CFU. These relatively large doses of vaccine induce diarrhea or vomiting in 15%–35% of volunteers. However, five doses of an SmD *S. flexneri* 2a vaccine did elicit serotype-specific protection in Yugoslav field trials (MEL et al. 1965b, 1971).

Trials of SmD vaccines in institutional populations in the United States have shown varying degrees of efficacy (reviewed in FORMAL and LEVINE 1984). These trials also indicated that streptomycin-independent revertants were shed by vaccinees, but these revertants were avirulent in the Sereny test. Since the revertants had lost the ability to propagate within the corneal epithelium of the guinea pig, the SmD vaccines used in these studies may have suffered other attenuating mutations in addition to streptomycin dependence. For example, a lyophilized SmD strain isolated 25 years ago has recently been shown to have a plasmid deletion that elimates the *ipaABCD* locus (A.B. HARTMAN, unpublished data). The instability of the SmD phenotype, the marginal protection observed in some United States field trials, and the requirement for multiple doses of $>10^{10}$ organisms have discouraged further work with these strains.

A second type of attenuated *Shigella* vaccine is characterized by both mutagen-induced purine auxotrophy (Pur^-) and spontaneous mutation in the RNA polymerase resulting in rifampin resistance (Rif) (LINDE et al. 1990). Purine

Table 1. *Shigella* vaccines with attenuating chromosomal mutations

Immunizing agent	Genotype	Phenotype	Reactogenicity in humans	Efficacy		References
				Monkey	Human	
SmD	*S. flexneri* 1, 2a, 3, 4 and *S. sonnei* with spontaneous mutations in *rpsL* ribosomal subunit gene resulting in streptomycin dependence	Limited in vivo multiplication, probably noninvasive	~15% at 5×10^{10}	NT	~50% protective in challenge studies and >80% protection in Yugoslav field trials using multiple doses	MEL et al. 1965; DUPONT et al. 1972
Pur^-/Rif	*S. flexneri* 2a and *S. sonnei* with mutagen-induced purine auxotrophy and spontaneous rifampin resistance	Limited in vivo multiplication, decreased invasion	~35% at 5×10^9	NT	None	LINDE et al. 1990; DENTCHEV et al. 1990
TSF-21	*S. flexneri* Y with spontaneous mutation in the *thyA* gene (thymine auxotrophy), and ultraviolet-induced mutation to temperature-sensitive growth (Ts^-)	Limited in vivo multiplication, invasion not tested	NT	100% protection after three doses	NT	AHMED et al. 1990
Sf1-114	*S. flexneri* Y with Tn*10* insertion in the *aroD* gene (PABA requirement)	Limited in vivo multiplication, invasive	NT[a]	100% protection after three doses	NT	LINDBERG et al. 1988; LINDBERG et al. 1990a

[a] An *aroD* deletion mutant (Sf1-124) elicited 10% reactogenicity at 1×10^9 (A.A. LINDSBERG, to be published)
NT, not tested

auxotrophy is not an attenuating characteristic because Pur^- strains retain the ability to evoke keratoconjunctivitis in the guinea pig eye. Nonetheless, some Pur^-/Rif double mutants are negative in the Sereny test, presumably as a result of an RNA polymerase mutation that alters a virulence-essential enzyme function. These double mutants retain the *Shigella* invasive phenotype as assessed by the infection of HeLa cells in vitro, but the level of invasion is decreased by up to 75% when compared to the wild-type parent strain. A Pur^-/Rif mutant of *S. flexneri* 2a was tested in a recent Bulgarian trial (DENTCHEV et al. 1990), and ingestion of approximately 5×10^9 CFU freshly harvested organisms was found to elicit reactions including tenesmus, meteorism, and diarrhea in 35% of vaccinees. Two of four vaccinees challenged with an ID_{50} of the homologous strain suffered dysentery. The reactogenicity of these vaccines at dosage levels that do not evoke a demonstrable protective immune response will probably discourage extensive human field testing.

A third type of attenuated vaccine is an *S. flexneri* Y strain with a *thyA* mutation isolated by thymine-trimethoprim selection. Unlike the purine auxotroph described above, this pyrimidine auxotroph is Sereny negative. However, the Thy^- mutant is genetically unstable, and a second mutation to temperature sensitivity (Ts) was induced by irradiation of this mutant with ultravidet light and by treatment with ampicillin at 39 °C. The resulting Thy^-/Ts vaccine candidate (designated TSF-21) was tested for safety and efficacy in the rabbits that had been conditioned by tetracycline treatment to reduce the normal flora, by neutralization of gastric acidity, and by opium-induced peristasis. When administered orally to the conditioned rabbits, TSF-21 is safe, and it protects against a subsequent *S. flexneri* Y challenge. Rhesus monkeys are also protected by three doses of 10^{11} CFU TSF-21 (AHMED et al. 1990). Although these results indicate that TSF-21 has efficacy in animal models, the strain must be grown at 30 °C, a temperature that is nonpermissive for expression of the invasive phenotype of *Shigella* species. Since it has not been demonstrated that the invasive phenotype is expressed while this strain is "coasting" at 37 °C in the intestinal lumen, it is probable that the TSF-21 vaccine is functionally noninvasive. Even though such noninvasive *Shigella* vaccines can elicit protection in primates and humans, they usually require large, multiple doses that limit their practical usefulness.

A fourth type of attenuated *Shigella* vaccine (Sfl-114) was constructed by transduction of the insertionally inactivated *aroD*: :Tn*10* gene of *E. coli* K-12 NK 5131 into *S. flexneri* serotype Y strain Sfl-1 (LINDBERG et al. 1988, 1990a). The *aroD* insertion inhibits the biosynthesis of aromatic metabolites such as chorismic acid, a precursor of *p*-aminobenzoic acid (PABA). The latter molecule is a biosynthetic component of tetrahydrafolate, a donor of the *N*-formyl group to *N*-formylmethionyl-tRNA. Primates acquire the folic acid precursor of tetrahydrofolate through dietary nutrients, and intermediary metabolism in monkeys and humans does not include a PABA biosynthetic pathway. Since shigellae cannot assimilate the folic acid that is available within mammalian tissues, survival of *aro* mutants that have invaded these tissues is limited by the gradual depletion of *N*-formylmethionyl-tRNA and inhibition of protein synthesis. For example, the growth of Sfl-114 within the cytoplasm of infected tissue culture cells ceases after four or five cell divisions. The vaccine is also negative in the Sereny test (LINDBERG et al. 1990b). Four doses of $2-3 \times 10^{10}$ CFU were safe and effective in protecting monkeys against a challenge with the Sfl-1 parent strain (LINDBERG et al. 1988). Most monkeys receiving four oral vaccinations of Sfl-114 responded with significant increases in serum IgA and IgG titer against serotype Y antigen (LINDBERG et al. 1990b).

An *aroD* deletion mutant of Sfl-114, designated Sfl-124, has been tested in Vietnamese and Swedish volunteers (A.A. LINDBERG et al., to be published). Although 10^9 CFU of the Sfl-124 vaccine was well tolerated in the Vietnamese population that had been exposed to endemic shigellosis, this dosage evoked transient diarrhea in approximately 10% of Swedish volunteers. Since Sf1-1, the parent strain of Sfl-124, evoked diarrhea·in 25% of volunteers in the United States who ingested 7×10^3 organisms, it can be surmised that the *aroD* deletion in Sfl-124 attenuates Sfl-1 by at least 100000-fold (LINDBERG and KARNELL 1990). However, the residual reactogenicity of Sfl-124 in immunologically naive volunteers suggests that aromatic mutants of *Shigella* species will not be safe when administered in dosages higher than 10^8 CFU. Lower doses of these genetically defined, nonreverting auxotrophs may prove to be both safe and protective in humans, and additional safety and efficacy studies of *aroD* shigella vaccine candidates are in progress.

3.2 *Shigella* Vaccines with Attenuating Plasmid Mutations

Table 2 summarizes some attributes of vaccines that are attenuated as a result of mutations in the *Shigella* virulence plasmid. The T_{32}Istrati strain was isolated in 1961 after 32 subcultures of *S. flexneri* 2a upon nutrient agar. During passage, random isolates were screened for loss of the Sereny-positive phenotype, and an isolate from the 32nd transfer proved to be a stable, avirulent mutant. Recent analysis of the genetic basis of attenuation in T_{32}Istrati indicates that the virulence plasmid in this strain has suffered deletions eliminating both the invasion region (*ipaBCDA* and *invA*) and the *icsA* gene that is necessary for intercellular bacterial spread. Since the Sereny-positive phenotype can be reconstituted in T_{32}Istrati by conjugal transfer of the invasion plasmid from *S. flexneri* serotype 5, it is evident that the safety and stability of T_{32}Istrati is dependent upon the noninvasive phenotype resulting from the plasmid deletions (M. VENKATESAN et al. 1991).

The T_{32}Istrati strain is probably the most extensively tested living, attenuated *Shigella* vaccine (MEITERT et al. 1984). Published results of field trials involving over 60000 individuals in Romania and over 5000 individuals in China indicate that T_{32}Istrati is safe even when given in large doses. Vaccination is associated with greater than 80% reduction in the incidence of *S. flexneri* infection. Unlike the SmD vaccines that induced serotype-specific immunity (MEL et al. 1965b,

Table 2. *Shigella* vaccines with attenuating plasmid mutations

Immunizing agent	Genotype	Phenotype	Reactogenicity in humans	Efficacy		References
				Monkey	Human	
T_{32}Istrati	*S. flexneri* 2a with spontaneous deletion in virulence plasmid	Noninvasive	$<1\%$ at 1×10^{11}	NT	$>80\%$ protection in Romanian and Chinese field trials using multiple doses of 10^{10} CFU	MEITERT et al. 1984; M. VENKATESAN et al. 1991
SC560 or SC570	*S. flexneri* 5 with deletion in *icsA* gene of virulence plasmid	Invasive, with limited intercellular spread	NT	100% protection after three doses	NT	SANSONETTI and ARONDEL 1989; P.J. SANSONETTI et al. this volume
SC5700	*S. flexneri* 5 with Tn*phoA* insertion in *icsA* gene of virulence plasmid and Tn*10* insertion in *iuc* chromosomal gene (aerobactin mutant)	Invasive, with limited intercellular spread	NT	~70% protection after three doses	NT	ibid

NT, not tested

1971), vaccination with T_{32}Istrati reportedly reduced the attack rates of heterologous *Shigella* serotypes (MEITERT et al. 1984). It should be noted that noninvasive strains of *S. flexneri* that are similar to T_{32}Istrati are taken up by the M cells overlying lymphoid follicles in rabbit ileal loops, but multiplication of these organisms within M cells is limited (WASSEF et al. 1989). Nonetheless, peroral immunization of rabbits with a noninvasive strain elicited a strong IgA memory response in secretions collected from chronically isolated ileal loops (KEREN et al. 1986). Presumably, the T_{32}Istrati vaccine functions in a similar manner to safely elicit a protective immune response.

The T_{32}Istrati field trials illustrate both the advantages and the disadvantages of a noninvasive mutant vaccine, i.e., it is safe when administered to individuals in endemic areas, but immunization requires five doses progressively escalating to 2.0×10^{11} CFU. This regimen is repeated every 6 months in order to maintain protection (MEITERT et al. 1984). T_{32}Istrati is produced and packaged by the Cantacuzino Institute in Bucharest as a suspension of 10^{11} CFU/ml in a synthetic preserving medium under paraffin oil. This preparation is stable at 4 °C for 60 days. In contrast to this refrigerated preparation, vaccines produced in the United States and Europe are routinely lyophilized and stored at −20° or −70 °C. Since recovery rates for these freeze-dried vaccines are usually in the 10%–20% range, vaccines requiring dosages of 10^{10} or 10^{11} CFU would be very expensive to produce. Therefore, repeated administration of large doses of reconstituted vaccine would be a daunting requirement for immunization programs to be carried out in developing countries.

More recent attempts to design attenuated *Shigella* vaccines by specific mutations in the virulence plasmids include *S. flexneri* 5 mutants SC560, SC570, and SC5700 (SANSONETTI and ARONDEL 1989) (Table 2). The plasnid genes encoding the invasive phenotype remain intact in these strains, and they retain the ability to invade tissue culture cells in vitro. Attenuation results from transduction of *icsA*: :Tn*pho*A into the virulence plasmid. The mutated *icsA* (BERNARDINI et al. 1989) or *virG* (MAKINO et al. 1986) gene is associated with the spread of intracellular shigellae to contiguous host cells in an infected tissue culture monolayer. As a result of the *icsA* mutation, both SC570 and SC5700 are unable to elicit keratoconjunctivitis in the Sereny test. Colonoscopy of rhesus monkeys vaccinated with *icsA* mutant SC560 (similar to SC570) revealed localized mucosal inflammation associated with isolated lymphoid follicles, but none of the generalized colitis that characterizes bacillary dysentery (P.J. SANSONETTI et al. this volume). These observations suggest that *icsA* mutants infect the FAE, but the subsequent spread of these organisms to the absorptive epithelium is limited.

In addition to the *icsA* mutation, strain SC5700 has also suffered a mutation in the chromosomal aerobactin gene *iucA* by *iuc*: :Tn*10* Pl*vir* transduction. Mutations in *iucA* have been shown to partially attenuate virulent shigellae in the Sereny test and the ligated rabbit ileal loop assay, but this attenuation can be overcome with a 100-fold increase of inoculum size (NASSIF et al. 1987). In rabbit loops, SC570 caused shortened villi with edema of the lamina propria, while SC5700 did not cause significant alterations of the general morphology of the ileal mucosa. Both strains evoked approximately one-half of the fluid volume that accumulated in loops injected with the *S. flexneri* 5 parent strain (SANSONETTI and ARONDEL 1989). Both the *icsA* mutant (SC560) and the *icsA-iucA* mutant (SC5700) elicit transient mucoid diarrhea in a minority of rhesus monkeys orally vaccinated with 10^9 CFU.

Three oral doses of either SC570 or SC5700 elicited significant serum immune response against the homologous *Shigella* somatic antigen and against *Shigella* invasion plasmid antigens. The vaccinated monkeys are significantly protected from a subsequent challenge with *S. flexneri* 5 (SANSONETTI and ARONDEL 1989). Nonetheless, the detectable reactogenicity of SC570 (or SC560) and SC5700 in preclinical animal tests suggests that *icsA* mutants of *Shigella* species may not be safe for humans when administered at a 10^9 dosage. Final determination of the safety and efficacy of *icsA* and *icsA-iucA* mutants awaits the construction of antibiotic sensitive versions of these mutants in an *S. flexneri* 2a background. These strains will be tested for safety in human volunteers.

4 *E. coli–S. flexneri* Hybrid Vaccines

The large intestine of primates normally contains a stable flora of facultative aerobes including *E. coli*. The capacity of this species to safely colonize the intestine, along with the ability to serve as a recipient for conjugal transfer of DNA from the closely related *Shigella* species, suggests that *E. coli* could be used as a carrier of *Shigella* antigens in hybrid vaccines. In the first test of this concept (LEVINE et al. 1977), a commensal strain of *E. coli* was used as a recipient in conjugal matings with an *S. flexneri* Hfr donor. The His and Met chromosomal markers were selected in sequential matings and the resulting *E. coli* hybrid, designated PGA142-1-5 (Table 3), expressed the serotype 2a somatic antigen. This noninvasive hybrid vaccine was completely safe when ingested by volunteers in three doses of 3×10^{10} CFU and it colonized for up to 2 weeks. However, PGA142-1-5 elicited increased serum antibody titers recognizing the 2a somatic antigen in fewer than 10% of vaccinees, and no protection was demonstrated upon subsequent challenge with virulent *S. flexneri* 2a. The failure of this hybrid to evoke a protective immune response suggests that colonization of the lumen of the bowel by a noninvasive *E. coli–Shigella* hybrid provides little stimulus to the GALT.

Approximately 5 years after the construction of PGA142-1-5, it was discovered that the genes encoding the enteroinvasive phenotype of *Shigella* species are carried on mobilizable plasmids. Conjugal transfer of the virulence plasmid from *S. flexneri* serotype 5 into *E. coli* K-12 confers a *Shigella*-like ability to invade tissue culture monolayers (SANSONETTI et al. 1983). This invasive

Table 3. *E. Coli–S. flexneri* hybrid vaccines

Immunizing agent	Genotype	Phenotype	Reactogenicity in humans	Efficacy		References
				Monkey	Human	
PGA142-1-15	*E. coli* 08 with His and Met markers from *S. flexneri* 2a	Noninvasive, expresses 2a somatic antigen	None at 3×10^{10}	NT	None	LEVINE et al. 1977
EC104 (CV61-1-2)	*E. coli* K-12 with His and Pro markers from *S. flexneri* 2a and virulence plasmid from *S. flexneri* 5[a]	Invasive with limited intercellular spread, expresses 2a somatic antigen	25% at 1×10^7	75% protection after two or three doses	NT	FORMAL et al. 1984
EcSf2a-1	*E. coli* K-12 with His, Pro, and Arg markers from *S. flexneri* 2a and virulence plasmid from *S. flexneri* 5[b]	Same as EC104	30% at 1×10^9	Same as EC104	No protection in challenge studies after three doses of 10^7 CFU	NEWLAND et al., in preparation K. KOTLOFF et al., in preparation
EcSf2a-2	EcSf2a-1 with deletion in the *aroD* gene	Same as EcSf2a-1 but with limited in vivo replication	None at 2×10^9	60% protection after three doses	50% protection in challenge studies after three doses of 10^9 CFU	ibid

[a] Tn*5* tags virulence plasmid with kanamycin resistance
[b] Tn*501* tags virulence plasmid with mercury resistance
NT, not tested

phenotype is closely associated with the ability of shigellae to infect the intestinal epithelium in orally challenged guinea pigs or rhesus monkeys (LABREC et al. 1964). In addition, invasive shigellae multiply within the M cells to a greater extent than noninvasive strains (WASSEF et al. 1989), and invasive shigellae are ingested more readily than noninvasive strains by cultured macrophages (CLERC et al. 1987). These observations suggest that *E. coli–Shigella* hybrids expressing the *Shigella* invasive phenotype might be more effective than the noninvasive PGA142-1-5 hybrid in delivering antigens to the GALT.

Attenuation of the invasive hybrids carrying the complete *Shigella* virulence plasmid is dependent upon differences between the *E. coli* and the *S. flexneri* chromosome. One *Shigella* virulence locus that is absent in *E. coli* K-12 chromosome is *iuc*, encoding aerobactin iron-binding siderophore system (DERBYSHIRE et al. 1989). Another difference is in the *purE*-linked *kcpA* (*k*eratoconjunctivitis *p*rovocation) locus that positively regulates the *Shigella* plasmid gene *icsA*. If the *purE* gene is not transferred from *S. flexneri* to an *E. coli* recipient carrying an *S. flexneri* invasion plasmid, spread of the invasive hybrid organism to contiguous cells in infected tissue culture monolayers rarely occurs (PAL et al. 1989). Since mutations in either *kcpA* (FORMAL et al. 1971) or *icsA* (*virG*) (MAKINO et al. 1986) attenuate *S. flexneri* in the keratoconjunctivitis Sereny test, intercellular spread of invasive organisms is apparently critical for the formation of lesions in the epithelial mucosa.

An invasive *E. coli* K-12 hybrid vaccine designated EC104 (FORMAL et al. 1984) (Table 3) was constructed by conjugal monilization of the *S. flexneri* 5 virulence plasmid followed by the transfer of the His and Pro markers from the chromosome of an *S. flexneri* 2a Hfr. The latter markers are linked to genes expressing the *S. flexneri* 2a group and type somatic antigen. EC104 evokes a slight inflammatory response in the ileal mucosa of ligated rabbit loops, but it does not cause fluid accumulation in ileal loops nor does it evoke keratoconjunctivitis in the Sereny test. EC104 was tested for safety and efficacy in rhesus monkeys by oral administration of two doses of 10^{11} organisms. This vaccine regimen was associated with diarrhea or dysentery in 15% of the monkeys, but the etiology of this reaction was questionable because *S. flexneri* 4 was isolated from the symptomatic monkeys. A second group of monkeys was inoculated with three doses of 1×10^{10} EC104, and no intestinal symptoms were observed. Approximately 30% of the vaccinated animals had significant rises in serum IgG recognizing the 2a somatic antigen and they were significantly protected against challenge with *S. flexneri* 2a (FORMAL et al. 1984).

The *S. flexneri* 5 invasion plasmid in EC104 is tagged with the Tn*5* kanamycin resistance transposon making this hybrid inappropriate for human testing. Therefore, a third *E. coli* K-12 hybrid vaccine was constructed by mobilization of an *S. flexneri* 5 virulence plasmid tagged with the Tn*501* mercury resistance tramsposon. *S. flexneri* 2a chromosomal regions linked to His, Pro, and Arg markers were then conjugally transferred into the invasive K-12 recipient. This construct, designated EcSf2a-1 (Table 3), was safe and protective when tested in rhesus monkeys. However, an oral dosage of 1×10^{9} CFU elicited diarrhea or dysentery in 30% of volunteers. Smaller doses (5×10^7 organisms) colonized the intestine of volunteers for up to 3 days in the absence of symptoms, but three doses of this regimen did not evoke protection against a subsequent challenge with *S. flexneri* 2a (D.A. HERRINGTON and M.M. LEVINE, unpublished data).

Variants that can spread from cell to cell in a tissue culture monolayer can be isolated at a low frequency from the EcSf2a-1 vaccine, and a pure culture of these variants can produce a hemorrhagic mucosa when injected into ligated rabbit ileal loops. This unexpected observation suggests that there is an unstable allele of the *Shigella kcpA* locus in the *E. coli* K-12 chromosome (PÁL et al. 1989). In order to further attenuate EcSf2a-1, the technique of LINDBERG and STOCKER (LINDBERG et al. 1988, 1990b) has been employed to generate a deletion mutation in the *aroD* gene. This mutation eliminates the ability of spreading variants to evoke hemorrhagic mucosal lesions in rabbit ileal loops and it also eliminates the residual fluid accumulation that was observed in loops injected with EcSf2a-1 (J.W. NEWLAND, T.L. HALE, and S.B. FORMAL, in preparation). The *aroD* mutant has been designated EcSf2a-2 (Table 3), and this vaccine has demonstrable safety and efficacy in rhesus monkeys (S.B. FORMAL, J.P. COGAN, and P.J. SNOY, unpublished data). Initial safety studies in humans revealed no diarrhea or dysentery when EcSf2a-2 was administered in three doses of 2×10^9 CFU within 7 days. Subsequent challenge of these volunteers with *S. flexneri* 2a indicated significant protection against bloody diarrhea, but the difference between vaccinees and unvaccinated controls did not reach statistical significance when all cases of diarrhea and fever were included in the assessment (K KOTLOFF and M.M. LEVINE, unpublished data). Nonetheless, the preclinical and clinical studies suggest that enteroinvasive *aroD E. coli-S. flexneri* hybrids may eventually prove to be useful *Shigella* vaccines, and further human studies are planned to assess the safety and efficacy of these vaccine constructs.

5 Ribosomal *Shigella* Vaccines

The previously discredited possibility of parenteral immunization against shigellosis has recently been revived by animal studies with ribosomal RNA vaccines prepared from *S. sonnei* or *S. flexneri.* The ribosomes were separated by polyethylene glycol precipitation of sonicated bacteria, and the immunizing component was shown to be a cytoplasmic O-polysaccharide (L-hapten) contaminant that is devoid of lipid A and KDO (LEVENSON and EGOROVA 1990). Two subcutaneous injections (200 μg per kilogram of body weight) of a ribosomal vaccine prepared from *S. sonnei* elicited secretory IgA recognizing the homologous somatic antigen in the saliva and bile in 60% of rhesus monkeys, and this vaccination regimen elicited at least 75% protection against diarrhea or dysentery (LEVENSON et al. 1988). The absence of reactogenicity when L-hapten-ribosomal vaccines are injected subcutaneously in monkeys or volunteers suggests that these vaccines could be used in humans, but efficacy studies in vaccinated volunteers have not been reported. It should also be noted that the O antigen content of seven *S. sonnei* rinosomal preparations varied by almost tenfold (LEVENSON and EGOROVA 1990), and this variability may preclude large-scale production of ribosomal vaccines using the current techniques.

6 Conclusions and Speculations

The approaches to vaccine development summarized in this chapter illustrate the difficulty of inducing even transient immunity against *Shigella* infection without evoking some clinical symptoms. The stability of attenuation is obviously critical to the safety of any *Shigella* vaccine, and spontaneous mutants such as SmD, Rif, *thyA*, and *kcpA* have a tendency to revert under the intense selective pressure in the lumen of the bowel. In contrast, site-directed deletion mutations such as *iuc*, *aroD*, and *icsA* provide more stable attenuation, but these mutations may not completely eliminate the reactogenicity of invasive vaccine candidates. Indeed, it may eventually become evident that any invasive enteric vaccine that delivers endotoxin to the GALT will be too reactogenic for human use. On the other hand, noninvasive mutants are relatively safe, but the multiple large doses that have been employed to demonstrate the efficacy of these vaccines present manufacturing and logistical problems.

Although balancing immunogenicity and reactogenicity in living, oral *Shigella* vaccines remains an elusive goal, application of new discoveries concerning the pathogenesis of *Shigella* species and the nature of the protective immune response against shigellosis will continue to generate many promising vaccine candidates. The constraint in testing these candidates is the absence of an animal model that can accurately predict either the safety or the efficacy of a vaccine in humans. The Sereny keratoconjunctivitis assay can be used to eliminate the more virulent vaccine candidates, but some Sereny-negative strains have been reactogenic when ingested by humans. On the other hand, the orally challenged monkey model may have some predictive value for vaccine efficacy, but this model is a poor predictor of vaccine safety. Identification of immune parameters that invariably correlate with protection in volunteers would be a key factor in accelerated *Shigella* vaccine development. This development would allow extensive phase I safety and immunogenicity trials of many vaccine candidates with selection of only the most promising strains for larger phase II safety trials and phase IIb challenge studies. Clarification of regulatory guidelines for the testing of genetically engineered bacteria on an outpatient basis will also be necessary for economical phase I and phase II testing of many current and future vaccine constructs.

Finally, it should be noted that living, oral bacterial vaccines mimicking the immune response elicited by natural *Shigella* infections may not prove to be the most effective immunoprophylactic approach. For example, *IpaBCD* and the *icsA* plasmid gene products are key determinants of the virulent phenotype of all members of the *Shigella* species, and immunoblots using convalescent serum from infected monkeys or humans indicate that these proteins are the immunodominant antigens eliciting antibody responses during *Shigella* infections (OAKS et al. 1986; DINARI et al. 1987). The absence of cross-reactive immunity among *Shigella* serotypes suggests that antibody recognizing these invasion plasmid antigens is not protective, but it is possible that there is serotype-specific heterogeneity in one or more Ipa genes that is not detectable by immunoblot techniques. These heterogeneous epitopes could conceivably be the basis of serotype-specific immunity that is currently attributed to the somatic antigen. Conversely, it is possible that there are conserved epitopes on Ipa proteins that are necessary for expression of the invasive phenotype in every *Shigella* species. These conserved epitopes may be sequestered in a way that evades T cell recognition during natural *Shigella* infections. If these hypothetical, conserved epitopes could be identified by genetic and/or by monoclonal antibody techniques, it is possible that a synthetic peptide vaccine could be designed to elicit *Shigella* genus-specific protective antibody that is not generated by natural shigella infections.

The ability of parenterally administered, L-hapten ribosomal vaccines to induce secretory IgA recognizing *Shigella* somatic antigen (LEVENSON et al. 1988, LEVENSON and EGOROVA 1990) has suggested that parenteral *Shigella* vaccines should receive renewed consideration. For example, detoxified LPS preparations consisting of O-polysaccharide conjugated to T cell-dependent protein carriers will be tested as parenteral *Shigella* vaccines in the near future (ROBBINS and SCHNEERSON 1990). In addition, noval methods of delivery of somatic antigen to the intestinal mucosa may prove to be safer and more effective than living, oral *Shigella* vaccines. For example, biodegradable poly (LD-lactide-coglycolide) microspheres (ELDRIDGE et al. 1989) incorporating *Shigella* LPS or Ipa proteins will soon be tested as an antigen delivery system for the GALT (S.B. FORMAL, personal communication). Given the uncertain utility of living, oral *Shigella* vaccines, alternative approaches such as synthetic peptides, glycoconjugates, and adjuvant delivery systems should receive continued attention.

Acknowledgements. We thank S.B. Formal and J.W. Newland of the Walter Reed Army Institute of Research, M.M. Levine and K. Kotloff of the Center for Vaccine Development of the University of Maryland School of Medicine, P.J. Sansonetti of the Institute Pasteur, and J.B. Robbins of the National Institutes of Health (USA) for the contribution of unpublished data.

References

Ahmed ZU, Mahfuzur RS, Sack DA (1990) Protection of adult rabbits and monkeys from lethal shigellosis by oral immunization with a thymine-requiring and temperature-sensitive mutant of *Shigella flexneri* Y. Vaccine 8: 153–158

Bennish ML, Harris JR, Wojtyniak BJ, Struelens M (1990) Death in shigellosis: incidence and risk factors in hospitalized patients. J Infect Dis 161: 500–506

Bernardini ML, Mounier J, D'Hauteville H, Coquis-Rondon M, Sansonetti PJ (1989) Identification of *icsA*, a plasmid locus of *Shigella flexneri* that governs bacterial intra- and intercellular spread through interaction with F-actin. Proc Natl Acad Sci USA 86: 3867–3871

Brandtzaeg P (1989) Overview of the mucosal immune system. In: Mestecky J, McGhee JR (eds) New strategies for oral immunization. Springer, Berlin Heidelberg New York, pp 13–25 (Current topics in microbiology and immunology, vol 146)

Clerc PL, Ryter A, Mounier J, Sansonetti PJ (1987) Plasmid-mediated early killing of eucaryotic cells by *Shigella flexneri* as studied by infection of J774 macrophages. Infect Immun 55: 521–527

Cohen D, Green MS, Block C, Rouach T, Ofek I (1988) Serum antibodies to lipopolysaccharide and natural immunity to shigellosis in an Israeli military population. J Infect Dis 157: 1068–1071

Czerkinsky C, Nilsson L-A, Hygren H, Ouchterlony O, Tarkowski AA (1983) A solid-phase enzyme-linked immunospot (ELISPOT) assay for enumeration of specific antibody-secreting cells. J Immunol Methods 65: 109–121

Dentchev V, Marinova S, Vassilev T, Bratoyeva M, Linde K (1990) Live *Shigella flexneri* 2a and *Shigella sonnei* I vaccine candidate strains with two attenuating markers. II. Preliminary results of vaccination of adult volunteers and children aged 2–17 years. Vaccine 8: 30–34

Derbyshire P, Stevenson P, Griffiths E, Roberts M, Williams P, Hale TL, Formal SB (1989) Expression in *Escherichia coli* K-12 of the 76 000-Dalton iron-regulated outer membrane protein of *Shigella flexneri* confers sensitivity to cloacin DF13 in the absence of *Shigella* O antigen. Infect Immun 57: 2794–2798

Dinari G, Hale TL, Austin SW, Formal SB (1987) Local and systemic antibody responses to *Shigella* infection in rhesus monkeys. J Infect Dis 155: 1065–1069

Dupont HL, Gangarosa EJ, Reller LB, Woodward WE, Armstrong RW, Hammond J, Glaser K, Morris GK (1970) Shigellosis in custodial institutions. Am J Epidemiol 92: 172–179

Dupont HL, Hornick RB, Snyder MJ, Libonati JP, Formal SB, Gangarosa EJ (1972) Immunity in shigellosis. II. Protection induced by oral live vaccine or primary infection. J Infect Dis 125: 12–16

Eldridge JH, Gilley RM, Staas JK, Moldoveanu Z, Meulbroek JA, Tice TR (1989) Biodegradable microspheres: vaccine delivery system for oral immunization. Springer, Berlin Heidelberg New York, pp 59–66 (Current topics in microbiology and immunology, vol 146)

Formal SB, Levine MM (1984) Shigellosis. In: Germanier R (ed) Bacterial vaccines. Academic, New York, pp 167–186

Formal SB, LaBrec EH, Palmer A, Falkow S (1965) Protection of monkeys against experimental shigellosis with attenuated vaccines. J Bacteriol 20: 63–68

Formal SB, Maenza RM, Austin S, LaBrec EH (1967) Failure of parenteral vaccines to protect monkeys against experimental shigellosis. Proc Soc Exp Biol Med 25: 347–349

Formal SB, Gemski P, Baron LS, LaBrec EH (1971) A chromosomal locus which controls the ability of *Shigella flexneri* to evoke keratoconjunctivitis. Infect Immun 3: 73–79

Formal SB, Hale TL, Kapfer C, Cogan JP, Snoy PJ, Chung R, Wingfield ME, Elisberg BL, Baron LS (1984) Oral vaccination of monkeys with an invasive *Escherichia coli* K-12 hybrid expressing *Shigella flexneri* 2a somatic antigen. Infect Immun 46: 465–469

Formal SB, Oaks EV, Olsen RE, Wingfield-Eggleston M, Snoy PJ, Cogan JP (1991) The effect of prior infection with virulent *Shigella flexneri* 2a on the resistance of monkeys to subsequent infection with *Shigella sonnei*. J Infect Dis 164: 533–537

Fubara ES, Freter R (1973) Protection against enteric bacterial infection by secretory IgA antibodies. J Immunol 111: 395–400

Gangarosa EJ, Perera DR, Mata LJ, Mendizábal-Morris C, Guzmán G, Reller LB (1969) Epidemic Shiga bacillus dysentery in Central America. II. Epidemiologic studies in 1969. J Infect Dis 122: 181–190

Hale TL, Bonventre PF (1979) *Shigella* infection of Henle intestinal epithelial cells: role of the bacterium. Infect Immun 24: 879–886

Herrington DA, Van de Verg L, Formal SB, Hale TL, Tall BD, Cryz SJ, Tramont EC, Levine MM (1990) Studies in volunteers to evaluate candidate *Shigella* vaccines: further experience with a bivalent *Salmonella typhi-Shigella sonnei* vaccine and protection conferred by previous *Shigella sonnei* disease. Vaccine 8: 353–357

Higgins AR, Floyd TM, Kadar MA (1955) Studies in shigellosis III. A controlled evaluation of a monovalent *Shigella* vaccine in a highly endemic environment. Am J Trop Med Hyg 4: 281–288

Keren DF, Holt PS, Collins HH, Gemski P, Formal SB (1978) The role of Peyer's patches in the local immune response of rabbit ileum to live bacteria. J Immunol 120: 1892–1898

Keren DF, Scott PJ, McDonald RA, Kern SE (1983) Local IgA memory response to bacterial antigens. Ann NY Acad Sci 409: 734–744

Keren DF, McDonald RA, Formal SB (1986) Secretory immunoglobulin A response following peroral priming and challenge with *Shigella flexneri* lacking the 140-megadalton virulence plasmid. Infect Immun 54: 920–923

Keren DF, Brown JE, McDonald RA, Wassef JS (1989) Secretory immunoglobulin A response to shiga toxin in rabbits: kinetics of the initial mucosal immune response and inhibition of toxicity in vitro and in vivo. Infect Immun 57: 1885–1889

LaBrec EH, Schneider H, Magnani TJ, Formal SB (1964) Epithelial cell penetration as an essential step in the pathogenesis of bacillary dysentery. J Bacteriol 88: 1503–1518

Levenson VI, Egorova TP (1990) Polysaccharide nature of O' antigen in protective ribosomal preparations from *Shigella*: experimental evidence and implications for the ribsomal vaccine concept. Res Microbiol 141: 707–720

Levenson VI, Chernokhvostova EV, Lybinskaya MM, Salamatova SA, Dzhikidze EK, Stasilevitch ZK (1988) Parenteral immunization with *Shigella* ribosomal vaccine elicits local IgA response and primes for mucosal memory. Arch Allergy Appl Immunol 87: 25–31

Levine MM, Woodward WE, Formal SB, Gemski P, DuPont HL, Hornick RB, Snyder MJ (1977) Studies with a new generation of oral attenuated shigella vaccine: *Escherichia coli* bearing surface antigens of *Shigella flexneri*. J Infect Dis 136: 577–582

Lindberg AA, Kärnell A, Stocker BAD, Katakura S, Sweiha H, Reinholt FP (1988) Development of an auxotrophic oral live *Shigella flexneri* vaccine. Vaccine 6: 146–150

Lindberg AA, Kärnell A, Pál T, Sweiha H, Hultenby K, Stocker BAD (1990a) Construction of an auxotropic *Shigella flexneri* strain for use as a live vaccine. Microb Pathog 8: 433–440

Lindberg AA, Kärnell A, Stocker BAD (1990b) Aromatic-dependent *Shigella* strains as live oral vaccines. In: Woodrow GC, Levine MM (eds) New generation vaccines. Marcel Dekker, New York, pp 677–687

Linde K, Dentchev V, Bondarenko V, Marinova S, Randhagen B, Bratoyeva M, Ṫesvetanov Y, Romanova Y (1990) Live *Shigella flexneri* 2a and *Shigella sonnei* I vaccine candidate strains with two attenuating markers. I. Construction of vaccine candidate strains with retained invasiveness but reduced intracellular multiplication. Vaccine 8: 25–29

Makino S, Sasakawa C, Kamata K, Kurata T, Yoshikawa M (1986) A genetic determinant required for continuous reinfection of adjacent cells on large plasmid in *Shigella flexneri* 2a. Cell 46: 551–555

Meitert T, Pencu E, Ciudin L, Tonciu M (1984) Vaccine strain *Sh. flexneri* T_{32}-ISTRATI. Studies in animals and volunteers. Antidysentery immunoprophylaxis and immunotherapy by live vaccine VADIZEN (*Sh. flexneri* T_{32}ISTRATI). Arch Roum Path Exp Microbiol 43: 251–278

Mel DM, PaPo RG, Terzin AL, Vuksic L (1965a) Studies on vaccination against bacillary dysentery. II. Safety tests and reactogenicity studies on a live dysentery vaccine intended for use in field trials. Bull WHO 32: 637–645

Mel DM, Terzin AL, Vuksic L (1965b) Studies on vaccination against bacillary dysentery. 3. Effective oral immunization against *Shigella flexneri* 2a in a field trial. Bull WHO 32: 647–655

Mel D, Gangarosa EJ, Radovanovic ML, Arsic BL, Litvinjenko S (1971) Studies on vaccination against bacillary dysentery. 6. Protection of children by oral immunization with streptomycin-dependent *Shigella* strains. Bull WHO 45: 457–464

Nassif X, Maxzert MC, Mounier J, Sansonetti PJ (1987) Evaluation with an *inc*: : Tn*10* mutant of the role of aerobactin production in the virulence of *Shigella flexneri*. Infect Immunol 55: 1963–1969

Oaks EV, Hale TL, Formal SB (1986) Serum immune response to *Shigella* protein antigens in rhesus monkeys and humans infected with *Shigella* spp. Infect Immun 53: 57–63

Owen RL (1977) Sequential uptake of horseradish peroxidase by lymphoid follicle epithelium of Peyer's patches in the normal unobstructed mouse intestine: an ultrastructural study. Gastroenterology 72: 440–451

Pál T, Newland JW, Tall BD, Formal SB, Hale TL (1989) Intracellular spread of *Shigella flexneri* associated with the *kcpA* locus and a 140-kilodalton protein. Infect Immun 57: 477–486

Reed WP (1975) Serum factors capable of opsonizing *Shigella* for phagocytosis by polymorphonuclear neutrophils. Immunology 28: 1051–1059

Robbins JB, Schneerson R (1990) Polysaccharide-protein conjugates: a new generation of vaccines. J Infect Dis 161: 821–832

Rosner AJ, Keren DF (1984) Demonstration of M-cells in the specialized follicle-associated epithelium overlying isolated follicles in the gut. J Leukocyte Biol 35: 397–404

Roux ME, McWilliams M, Phillips-Quagliata JM, Lamm ME (1981) Differentiation pathways of Peyer's patch precursors of IgA plasma cells in the secretory immune system. Cell Immunol 61: 141–153
Sansonetti PJ, Arondel J (1989) Construction and evaluation of a double mutant of *Shigella flexneri* as a candidate for oral vaccination against shigellosis. Vaccine 7: 443–450
Sansonetti PJ, Arondel J, Fontaine A, d'Hauteville H, Bernardini ML, Sansonetti et al. (1991) *ompB* (osmo-regulation) and *icsA* (cell to cell spread) mutants of *Shigella flexneri* are vaccine candidates and probes to study the pathogenesis of shigellosis vaccine.
Sansonetti PJ, Hale TL, Dammin GJ, Kapfer C, Collins HH, Formal SB (1983) Alterations in the pathogenicity of *Escherichia coli* K-12 after transfer of plasmid and chromosomal genes from *Shigella flexneri*. Infect Immun 39: 1392–1402
Smith MW, James PS, Tivey DR (1987) M cell numbers increase after transfer of SPF mice to a normal animal house environment. Am J Pathol 128: 385–389
Sneller MC, Strober W (1986) M cells and host disease. J Infect Dis 154: 737–741
Tagliabue A, Boraschi D, Villa L, Keren DF, Lowell GH, Rappuoli R, Nencioni L (1984) IgA-dependent cell-mediated activity against enteropathogenic bacteria: distribution, specificity, and characterization of the effector cells. J Immunol 133: 988–992
Trejosiewicz LK, Malizia G, Badr-el-Din S, Smart CJ, Oakes DJ, Southgate J, Howdle PD, Janossy G (1987) T cell and mononuclear phagocyte populations of human small and large intestine. Adv Exp Med Biol 216A: 465–473
Tseng J (1984) A population of resting IgM-IgD double-bearing lymphocytes in Peyer's patches: the major precursor cells for IgA plasma cells in the gut lamina propria. J Immunol 132: 2730–2734
Venkatesan M, Fernandez-Prada C, Buysee JM, Formal SB, Hale TL (1991) Virulence phenotype and genetic characteristics of the T_{32} *Shigella flexneri* 2a vaccine strain. Vaccine 9: 358–363
Wassaf JS, Keren DF, Mallous JL (1989) Role of M cells in initial antigen uptake and in ulcer formation in the rabbit intestinal loop model of shigellosis. Infect Immun 57: 858–863

Subject Index

Current Topics in Microbiology and Immunology

Volumes published since 1988 (and still available)

Vol. 136: **Hobom, Gerd; Rott, Rudolf (Ed.):** The Molecular Biology of Bacterial Virus Systems. 1988. 20 figs. VII, 90 pp. ISBN 3-540-18513-5

Vol. 137: **Mock, Beverly; Potter, Michael (Ed.):** Genetics of Immunological Diseases. 1988. 88 figs. XI, 335 pp. ISBN 3-540-19253-0

Vol. 138: **Goebel, Werner (Ed.):** Intracellular Bacteria. 1988. 18 figs. IX, 179 pp. ISBN 3-540-50001-4

Vol. 139: **Clarke, Adrienne E.; Wilson, Ian A. (Ed.):** Carbohydrate-Protein Interaction. 1988. 35 figs. IX, 152 pp. ISBN 3-540-19378-2

Vol. 140: **Podack, Eckhard R. (Ed.):** Cytotoxic Effector Mechanisms. 1989. 24 figs. VIII, 126 pp. ISBN 3-540-50057-X

Vol. 141: **Potter, Michael; Melchers, Fritz (Ed.):** Mechanisms in B-Cell Neoplasia 1988. Workshop at the National Cancer Institute, National Institutes of Health, Bethesda, MD, USA, March 23–25, 1988. 1988. 122 figs. XIV, 340 pp. ISBN 3-540-50212-2

Vol. 142: **Schüpach, Jörg:** Human Retrovirology. Facts and Concepts. 1989. 24 figs. 115 pp. ISBN 3-540-50455-9

Vol. 143: **Haase, Ashley T.; Oldstone Michael B. A. (Ed.):** In Situ Hybridization 1989. 22 figs. XII, 90 pp. ISBN 3-540-50761-2

Vol. 144: **Knippers, Rolf; Levine, A. J. (Ed.):** Transforming. Proteins of DNA Tumor Viruses. 1989. 85 figs. XIV, 300 pp. ISBN 3-540-50909-7

Vol. 145: **Oldstone, Michael B. A. (Ed.):** Molecular Mimicry. Cross-Reactivity between Microbes and Host Proteins as a Cause of Autoimmunity. 1989. 28 figs. VII, 141 pp. ISBN 3-540-50929-1

Vol. 146: **Mestecky, Jiri; McGhee, Jerry (Ed.):** New Strategies for Oral Immunization. International Symposium at the University of Alabama at Birmingham and Molecular Engineering Associates, Inc. Birmingham, AL, USA, March 21–22, 1988. 1989. 22 figs. IX, 237 pp. ISBN 3-540-50841-4

Vol. 147: **Vogt, Peter K. (Ed.):** Oncogenes. Selected Reviews. 1989. 8 figs. VII, 172 pp. ISBN 3-540-51050-8

Vol. 148: **Vogt, Peter K. (Ed.):** Oncogenes and Retroviruses. Selected Reviews. 1989. XII, 134 pp. ISBN 3-540-51051-6

Vol. 149: **Shen-Ong, Grace L. C.; Potter, Michael; Copeland, Neal G. (Ed.):** Mechanisms in Myeloid Tumorigenesis. Workshop at the National Cancer Institute, National Institutes of Health, Bethesda, MD, USA, March 22, 1988. 1989. 42 figs. X, 172 pp. ISBN 3-540-50968-2

Vol. 150: **Jann, Klaus; Jann, Barbara (Ed.):** Bacterial Capsules. 1989. 33 figs. XII, 176 pp. ISBN 3-540-51049-4

Vol. 151: **Jann, Klaus; Jann, Barbara (Ed.):** Bacterial Adhesins. 1990. 23 figs. XII, 192 pp. ISBN 3-540-51052-4

Vol. 152: **Bosma, Melvin J.; Phillips, Robert A.; Schuler, Walter (Ed.):** The Scid Mouse. Characterization and Potential Uses. EMBO Workshop held at the Basel Institute for Immunology, Basel, Switzerland, February 20–22, 1989. 1989. 72 figs. XII, 263 pp. ISBN 3-540-51512-7

Vol. 153: **Lambris, John D. (Ed.):** The Third Component of Complement. Chemistry and Biology. 1989. 38 figs. X, 251 pp. ISBN 3-540-51513-5

Vol. 154: **McDougall, James K. (Ed.):** Cytomegaloviruses. 1990. 58 figs. IX, 286 pp. ISBN 3-540-51514-3

Vol. 155: **Kaufmann, Stefan H. E. (Ed.):** T-Cell Paradigms in Parasitic and Bacterial Infections. 1990. 24 figs. IX, 162 pp. ISBN 3-540-51515-1

Vol. 156: **Dyrberg, Thomas (Ed.):** The Role of Viruses and the Immune System in Diabetes Mellitus. 1990. 15 figs. XI, 142 pp. ISBN 3-540-51918-1

Vol. 157: **Swanstrom, Ronald; Vogt, Peter K. (Ed.):** Retroviruses. Strategies of Replication. 1990. 40 figs. XII, 260 pp. ISBN 3-540-51895-9

Vol. 158: **Muzyczka, Nicholas (Ed.):** Viral Expression Vectors. 1992. 20 figs. IX, 176 pp. ISBN 3-540-52431-2

Vol. 159: **Gray, David; Sprent, Jonathan (Ed.):** Immunological Memory. 1990. 38 figs. XII, 156 pp. ISBN 3-540-51921-1

Vol. 160: **Oldstone, Michael B. A.; Koprowski, Hilary (Ed.):** Retrovirus Infections of the Nervous System. 1990. 16 figs. XII, 176 pp. ISBN 3-540-51939-4

Vol. 161: **Racaniello, Vincent R. (Ed.):** Picornaviruses. 1990. 12 figs. X, 194 pp. ISBN 3-540-52429-0

Vol. 162: **Roy, Polly; Gorman, Barry M. (Ed.):** Bluetongue Viruses. 1990. 37 figs. X, 200 pp. ISBN 3-540-51922-X

Vol. 163: **Turner, Peter C.; Moyer, Richard W. (Ed.):** Poxviruses. 1990. 23 figs. X, 210 pp. ISBN 3-540-52430-4

Vol. 164: **Bækkeskov, Steinnun; Hansen, Bruno (Ed.):** Human Diabetes. 1990. 9 figs. X, 198 pp. ISBN 3-540-52652-8

Vol. 165: **Bothwell, Mark (Ed.):** Neuronal Growth Factors. 1991. 14 figs. IX, 173 pp. ISBN 3-540-52654-4

Vol. 166: **Potter, Michael; Melchers, Fritz (Ed.):** Mechanisms in B-Cell Neoplasia 1990. 143 figs. XIX, 380 pp. ISBN 3-540-52886-5

Vol. 167: **Kaufmann, Stefan H. E. (Ed.):** Heat Shock Proteins and Immune Response. 1991. 18 figs. IX, 214 pp. ISBN 3-540-52857-1

Vol. 168: **Mason, William S.; Seeger, Christoph (Ed.):** Hepadnaviruses. Molecular Biology and Pathogenesis. 1991. 21 figs. X, 206 pp. ISBN 3-540-53060-6

Vol. 169: **Kolakofsky, Daniel (Ed.):** Bunyaviridae. 1991. 34 figs. X, 256 pp. ISBN 3-540-53061-4

Vol. 170: **Compans, Richard W. (Ed.):** Protein Traffic in Eukaryotic Cells. Selected Reviews. 1991. 14 figs. X, 186 pp. ISBN 3-540-53631-0

Vol. 171: **Kung, Hsing-Jien; Vogt, Peter K. (Eds.):** Retroviral Insertion and Oncogene Activation. 1991. 18 figs. X, 179 pp. ISBN 3-540-53857-7

Vol. 172: **Chesebro, Bruce W. (Ed.):** Transmissible Spongiform Encephalopathies. 1991. 48 figs. X, 288 pp. ISBN 3-540-53883-6

Vol. 173: **Pfeffer, Klaus; Heeg, Klaus; Wagner, Hermann; Riethmüller, Gert (Ed.):** Function and Specificity of γ/δ T Cells. 1991. 41 figs. XII, 296 pp. ISBN 3-540-53781-3

Vol. 174: **Fleischer, Bernhard; Sjögren, Hans Olov (Eds.):** Superantigens. 1991. 13 figs. IX, 137 pp. ISBN 3-540-54205-1

Vol. 175: **Aktories, Klaus (Ed.):** ADP-Ribosylating Toxins. 1992. 23 figs. IX, 148 pp. ISBN 3-540-54598-0

Vol. 176: **Holland, John J. (Ed.):** Genetic Diversity of RNA Viruses. 1992. 34 figs. IX, 226 pp. ISBN 3-540-54652-9

Vol. 177: **Müller-Sieburg, Christa; Torok-Storb, Beverly; Visser, Jan; Storb, Rainer (Eds.):** Hematopoietic Stem Cells. 1992. 18 figs. XIII, 143 pp. ISBN 3-540-54531-X

Vol. 178: **Parker, Charles J. (Ed.):** Membrane Defenses Against Attack by Complement and Perforins. 1992. 26 figs. VIII, 188 pp. ISBN 3-540-54653-7

Vol. 179: **Rouse, Barry T. (Ed.):** Herpes Simplex Virus. 1992. 9 figs. X, 180 pp. ISBN 3-540-55066-6

MIX
Papier aus verantwortungsvollen Quellen
Paper from responsible sources
FSC® C105338

If you have any concerns about our products,
you can contact us on
ProductSafety@springernature.com

In case Publisher is established outside the EU,
the EU authorized representative is:
Springer Nature Customer Service Center GmbH
Europaplatz 3, 69115 Heidelberg, Germany

Printed by Libri Plureos GmbH
in Hamburg, Germany